SLEEP WHEN YOU'RE DEAD

Reviews for *KILLING STATE*

'A high-octane plot that centres around the dark heart of British political power. A great debut.'
SUNDAY TIMES

'A terrific future-shock thriller full of pace, tension, character, and emotion.'
LEE CHILD

'Thought-provoking, pacy and thrilling.'
SUNDAY MIRROR

'Fast, sharply written, clever and intense – but with tenderness and great characterisation too.'
Jeremy Vine, *BBC Radio 2*

Reviews for *CURSE THE DAY*

'Starts off like a fired bullet and never lets up. A sheer delight.'
DAVID BALDACCI

'A masterful storyteller.'
LJ ROSS

'A rip-roaring road trip into the dark heart of a corrupt, cynical British establishment.'
FINANCIAL TIMES

'A constantly surprising, heart-felt, desperately exciting super-thriller.'
ROB PARKER

BY JUDE O'REILLY

Wife in the North
A Year of Doing Good

Michael North Series

Killing State
Curse the Day
Sleep When You're Dead

SLEEP WHEN YOU'RE DEAD

JUDE O'REILLY

An Aries Book

First published in the UK in 2022 by Head of Zeus Ltd,
part of Bloomsbury Publishing Plc

9 7 5 3 1 2 4 6 8

A catalogue record for this book is available from
the British Library.

ISBN (HB): 9781801109468
ISBN (XTPB): 9781801109475
ISBN (E): 9781801109444

Typeset by Divaddict Publishing Solutions

Printed and bound in Great Britain by
CPI Group (UK) Ltd, Croydon CR0 4YY

MIX
Paper from
responsible sources
FSC® C171272

Head of Zeus Ltd
First Floor East
5–8 Hardwick Street
London EC1R 4RG
WWW.HEADOFZEUS.COM

In memory of C.

1

London

His heart was going to burst if he had to keep running. His lungs ripping and shredding as he forced too much air into them. Each jolt of the pavement shuddering through the rubber soles of his boots and up his spine, breaking into his skull and rattling the bullet inside it.

The man in front, swerving through the Londoners and tourists, risked a look behind, and Michael North forced himself to move faster. He had to catch this maniac before he used one of those butcher's knives gaffer-taped to his hands. Screams now as the evening crowds around the Tower of London began to realize what was going on. Irritation at the rudeness of being shoved to one side replaced by a stark and primal terror. Or was the knifeman already cutting victims into pieces? *Don't let him have killed some innocent*, North had time to pray, before arterial blood sprayed into the air ahead of him and the piercing shriek of a young girl cut through the screams, then another and another, until the hurting of strangers was all he could hear.

Anger then, and North stopped caring about his own pain, running cold and clean into it, using it as fuel to power himself onwards and onto Tower Bridge. The government agent could take pain, had before, and would again. The knifeman ahead of him, he powered across the road, narrowly avoiding a black cab, a London bus – the squeal of brakes – hurdling the crash barrier. He'd barely been aware of the distant sirens, the red lights, the metal gate stopping traffic as he shoved his way through the mass of people running the other way. The jostle and push of the stampeding crowd coming in his direction, everyone running, falling onto their hands and knees, stumbling to their feet again, helping each other up, knocking each other down. Their desperation to get away hampering North's own attempts to shoulder a path through.

In his peripheral vision, glass and lights, and the sense of something bobbing in the water beyond the fortress, a tall-masted boat, if he had to guess. North wasn't wasting time that he didn't have by checking. Everything focused on the whereabouts of the knifeman. The shrieks and screams getting louder as North closed down the distance between himself and his target. The crowd ahead was blocking the knifeman's route. *Let the lunatic through*, North thought. *Don't let anyone else get hurt. He's mine.* And his frustration exploded into something dark and violent.

The cast-iron gate blocking the pedestrian way over the bridge was closed. Undeterred, the knifeman leaped for it, clung, then scrambled over. He kept running.

North followed.

With a lurch, under his feet the bascule of the bridge began to lift, and it felt like hope. This was the moment – the knifeman would have to stop, and when he did North was going to kill him where he stood with his own damn knives. And, as he thought it, his eyes locked onto his target and the world shrank away to just the two of them. The running man was tall. A shabby workwear jacket, a glimpse of a beak nose – the strangely elongated arms. Time slowing. Tick tock. And

North's heart slowed, the pulse in his ears and the rise and fall of his breath – in and out – muffling the sounds of chaos around them.

The deck shifted and, instinctively, North leaned forward. He kept moving, hard and fast, and ahead of him the knifeman turned again, further around this time. A grimace as he spotted North behind. A hand half raised, beckoning onwards, as if he didn't want to be alone. North felt his mouth open as he yelled something. Didn't take in what he himself was shouting. *Stop!* he figured. But the man didn't seem like he was able to stop.

North was closing in on him now – his hand reaching, his fingers ready to grasp hold. But, almost as if he sensed the danger, the knifeman picked up his pace and North was forced to do the same.

The sensation of running uphill as North took the slope of the bridge faster than he'd thought possible. Surely, the guy wasn't going to do what it looked like he was going to do?

Was he? The running man would have to stop. A lick of pleasure at the anticipation of bringing the guy down.

But, instead of stopping, the knifeman disappeared out of view – he had leaped into the void. Jumping from the top of the lifting deck into nothingness was madness. North's rational brain knew that, just as he knew he shouldn't follow.

Was he already in the Thames? Was he dead? North was only seconds behind the other man. But they were seconds during which the deck of the bridge kept lifting. No, if there was one thing North shouldn't do, it was to make an impossible jump after a desperate lunatic. Couldn't. Wouldn't. Shouldn't make that jump. But even as the gap between one side and the other grew wider, North's body readied itself – a thousand tiny subconscious calculations tweaking every muscle, tendon and ligament.

It would be an act of stupidity, if not downright suicide, to follow a madman off an opening bascule bridge, in hope rather than expectation of landing on the other side. North would almost certainly die. He really shouldn't make the leap, he knew, even as his lungs sucked down more air in preparation for it, and his mind tried and failed to persuade him to slow down. His body was breakable. Shouldn't his body be listening to what reason was saying? But still, North kept building speed, running faster, harder. Pumped. Because his priority was not and never would be his own breakable body, it was the predator with knives instead of hands. The predator who – if he wasn't in the Thames or dead – could already be slipping and sliding down the far side of the bridge to scythe and carve his bloody way through innocent Londoners, and North couldn't have that. Anything was better than that. Wasn't it?

The climb was getting steeper.

No! he ordered himself. *Abort mission!* he told his body. Because making the jump was the act of a reckless lunatic. And he wasn't a reckless lunatic.

Was he?

Maybe so and maybe not. But the truth was North was a predator as much as any lone wolf with knives for hands. The rhythm of his pace was out of step with the rhythm of his heart, everything disconnected and out of sorts. And there was a moment when North wanted to laugh. Roadrunner. Himself as a kid watching a cartoon wolf running straight off a clifftop, his furry legs a blur as he tried to stay up, the long drop down into the canyon. Always the same. The cartoon wolf never learning from history – an endless plummet to his doom. And if ever there was a time to listen to reason rather than follow some primitive instinct, it was right this second. There would

4

be police on the other side of the bridge, he assured himself. It wasn't up to him alone to stop the bad guy and save the world. He had done everything he could. North gave himself permission to pull away from the chase, to stop before it was too late, but his body wouldn't stop moving. Wanting to stop and wanting not to stop. Because he *hadn't* done everything. Not yet anyway.

North knew his own folly even as his knee bent, taking the weight, powering up through the ground, forcing the kinetic energy through his glutes, hamstrings, quadriceps and calf muscles, the heel and ball of his foot, lifting off, scissoring through the cold spring air, reaching with every fibre for the other side of the bridge. The Thames below. Travelling through the chill and biting wind. Desperate for the hard landing of the bridge. Still reaching, grasping only sky, reaching further. Willing himself to touch the other side, which was still moving away from him.

Rueful – life sweeter with each and every passing second. Had he thrown what he had left of it away? The nothingness above and below him wanting him, claiming him.

But not giving up. He'd learn to fly if that's what it took.

The impact as his forehead and then his fingers found the edge of metal, and his body swung forwards, before smashing into the iron-girdered underbelly of the bridge with enough brutal force to knock what breath he had left clear out of him. Relief flooded his system. He wasn't falling. Then, a dazed kind of panic – he hadn't made it to safety. Instead, he hung from the lowest edge of the bascule, his arms nearly wrenched from their sockets, his grip clenched tight, the certainty of the waiting, churning Thames below.

His head rehearsed the manoeuvre his body would have to make a split second before his body took over again to pull himself onto the edge of the roadway. He heaved and straightened out his arms, brought first one leg up and then the other. The pavement under his knees the best thing he'd felt since forever.

His head throbbed with pain.

A shadow fell over him and with it came the rank smell of unwashed clothes and animal decay. He was too close to the edge, he realized, staring at the trainers that belonged to the shadow man. Cold wind on the nape of his neck and that immense waiting nothingness behind him – the fear of the infinite drop. The sound of the city in the distance, sirens and traffic, the slap and brown churn of the river below.

He had to stand up.

Some trick of the light or the blow to his head had turned everything North could see into black and white and shades of grey. The only real colour, the bloodied blades of the knives that dangled from the man's sleeves, their tips weeping fat crimson tears. North lifted his gaze, sure that even that movement was enough to take him over the edge. He met the black pools of the other man's eyes as he leaned over him, and he saw a dark kind of welcome there. North kept his own hands in view as he rose unsteadily to his feet, swaying, his heels still hanging over the edge, attempting to keep his weight forward. Bleary from the blow, he eased himself forward an inch, two, so that the soles of his feet made full contact with the road surface. He felt himself tip forwards and then backwards and forwards again. And the mouth of the other man widened into the kind of smile that told him the lunatic was delighted that North had made it across to his black and white world, because now North belonged to him.

The weight of a bony hand on his shoulder as the knifeman spoke into his ear. The accent was Scottish. 'Five... eight... two.'

It was a chant of sorts. The swing of a leather bag on a cord around the scrawny neck. 'Five... seven.' North tried to catch hold of their meaning, but the numbers meant nothing to him. He thought he caught his own name, but that was impossible. The man was a stranger. Bad breath, foetid, as if something had died inside. More numbers that he couldn't hold on to. North resisted pulling away from the stench, and the bulging cheekbone knocked against him, the stubble of the man's cheek prickling his own skin. 'We all have to die, Brother.' As the man spoke, North felt the left hand crawl to the top of his head, the blade catching and cutting as it went, before resting there as if in bloody benediction, all the time the scratchy voice growing quieter and more sinister, as if imparting a secret. 'One... two... two... six... zero... four. You'll be first in line...' The knifeman swung his other arm wide, the weapon gripped tight, the silver gaffer tape sodden with blood. '...Brother.' The final word a roar.

The knife began its arc; North turned his head and watched the blade moving through the air as if hypnotized.

North knew that he would never make old bones. And there were any number of ways a man in North's line of work could die. But, as the sun came out from behind a cloud, he chose not to die right here, right now, at the edge of the world and at the hands of a lunatic. No, this wasn't how it ended, he told himself, blocking the scarecrow's arm with his own forearm. Because he had a choice in this at least. Death was inevitable, but he could choose how he died. And using the palms of his scraped-about hands, North pushed himself away from the madman, then stepped backwards away from the edge of the bridge, out into the waiting air, the endless rushing drop down into the abyss, and the shocking embrace of the River Thames.

2

THREE MONTHS EARLIER

Glen Shiel

Laurence Sampson did not regard himself as a brave man. If anything, he knew himself to be a coward of the first order, which is why he was running away from trouble. He despised himself for it. But he wasn't willing to confront his bosses, military or commercial. He was repelled by their unblinking focus, by their single-minded attachment to the power to destroy, and by their willingness to use words such as peace when they meant war. They were patriots one and all and in the opinion of Laurie Sampson, a patriot could justify almost anything.

He himself was not a patriot, nor had he ever been. Laurie believed in two things. Science and his sister. Science would never confound you. He had dedicated his life to it and up to this moment, he had never regretted the sacrifices he had made for it. But it couldn't go on.

The road through the Highlands was almost deserted, but then it was the middle of the night. He turned up the air con as cold as it would go, before swigging another mouthful of a lukewarm energy drink in his cup holder, waiting for one or the other to jolt him into wakefulness.

Lucy, he thought. He would ring from Glasgow airport

before he got on his flight. He would tell her he was coming home and get her to write it down so she didn't forget. Anxiously, he tapped his top pocket to check his phone was in there, the ticket already downloaded. Whatever he was giving up by way of money or status, he would be able to look after Lucy better. She needed him and that came first. And if he was being honest with himself, going back home to London to care for Lucy gave him the excuse he needed to walk away from his work.

At first, they thought it was a ruse to boost his pay. But he didn't need more money. He had more than enough to pay for his needs and plenty to send to Lucy so she could live in comfort. He'd seen something shift in their eyes when they realized he couldn't be bought.

What then? they'd asked – their faces cold.

We thought you were happy. Aren't you happy?

So shallow – all that talk of happiness. So very American.

Laurie didn't think he had ever been happy. Unless you counted the stillness that swept over him sometimes when he was truly in the zone. When entire days and nights would pass without him realizing. Or again, when he finished a particularly difficult crossword puzzle. The Saturday one in *The New York Times*, for instance, had its moments. Or again, when he set a fiendish clue of his own in one of the puzzles he devised for Lucy.

She was his twin, he told them, the palms of his hands sweaty. And she was sick – early onset dementia – and she was getting worse. He'd had to avert his gaze so they didn't notice the nervous tic start up in his right eye. Yes, it was a dreadful disease. He didn't know how many good years she had left. But she needed him. He was sorry – he heard the feebleness in his own voice – he really had no choice in the matter.

9

And their faces had clouded over, till they were difficult to read, but what could they do, other than accept his resignation.

No, he was relieved he was leaving it all behind. It was all such a mess. What had he been thinking, to allow himself to get caught up in it? He should have known better. He glanced out of his window and up into the sky – so many stars. Was that a shooting star? Or a satellite? One of theirs, perhaps? Were they watching him right this second? He looked back to the winding road, then stamped on his brake full force, and the SUV slammed to a stop. The stag caught in his headlights as it crossed the road blinked in surprise, twitched its ears, then slowly, as if it had all the time in the world, moved off into the shadows and across to the other side. He watched it go, grateful for the unexpected wonder of it.

His shoulder sore from the seat belt, his breathing rapid and heart pounding, he began to laugh. This then was the happiness the Americans talked about so much. He saw the why of it. He was out and this was something to tell Lucy. The spreading antlers, he thought. He'd tell her how the stag held its head under the weight of the antlers, she'd like that. And what kind of crossword clue could he build around a stag? Maybe some reference to Sir Walter Scott?

She'd like that too, he thought, and then the full-beam headlights flooded his rear-view mirror and he knew, even before he felt the impact, that he'd been right to be frightened.

3

NOW

London

Michael North was not someone with time to waste. He'd felt better, but the hanging IV bag was empty – the antibiotics were in his system, which meant, as far as he was concerned, he was good to go. He shivered. Hours had passed but he was still cold from the Thames. Even so, bearing in mind he'd been forced to leap a chasm, plunge way too far and fast into a river, and almost drown, he was – all things considered – in rather a good mood. After all, it could have gone a lot worse – for him at least. He started to whistle as he ripped away the medical tape, then cursed as he tugged at the cannula tap spiked into the vein on his forearm and it caught in his flesh.

It was midnight and he was keen to be out of hospital. Ever since he'd taken a bullet to his brain in combat six years earlier – a bullet that was still lodged there – he'd hated hospitals with a vengeance. At the time, doctors described the bullet's trajectory and position – just short of the posterior parietal artery in the right temporo-parietal junction – as one in a million, and himself as 'freakishly lucky'. And it would appear the luck was holding. That's to say, he wasn't dead yet. Although, he could do without the smell of antiseptic and misery in his nose

– again. Nearly four hours in hospital was nearly four hours too long as far as he was concerned, and, impatient, he gripped the small tap more tightly as he pulled harder. As the needle emerged, a spurt of blood scythed through the air to fall away into a splatter of bright red blots against the white sheet. He closed his eyes against the memory of blood falling from the tips of the knives.

Edmund Hone, his one-eyed boss and head of the secretariat within MI5 known as the 'Friends of Cyclops', had arrived only minutes before. He now stood looking out of the window into the night, listening hard to whoever was on the other end of the line. He hadn't asked how North was feeling, merely nodded as he walked into the room, which North took to be a sign of approval that he hadn't been so inconsiderate as to die on the job, which would mean paperwork.

When Hone finally came off the phone, North knew he should ask about the madman on the bridge, but he wasn't sure if he wanted to know. The grim realization that he had to step off backwards into oblivion or die where he stood stayed with him – the sensation of plummeting through the air, the bone-shuddering impact as he crashed into the freezing cold water, the suck of the mud as his boots sank into the riverbed, which did its best to hold tight to him, and, when he did manage to free himself, the darkness and vicious tug of the current as it tried to rip him away from this life; he shuddered as he cycled through them and out the other side. It was as well he suffered from chronic insomnia; it cut down on the opportunity for nightmares. And while he was looking on the bright side, frankly, he'd been lucky to be pulled from the water as fast as he was.

'Gephyrophobia is a real thing, you know.' Fangfang Yu's small chin was balanced on her fist, her elbow pressing against her knee, her glittering Dr. Martens boots shedding gold over the folded blanket as she sat cross-legged at the bottom of his bed. She yawned. 'Fear of bridges. Sufferers will go miles out of their way to avoid crossing one.' The teenage geek kept her voice innocent, the Geordie accent soft and warm, but the eyes behind the heavy black-framed glasses were gleeful.

His friend Plug let out a snort of laughter. North had known Padraig 'Plug' Donne since their days in a young offender institution. It wasn't the first time his old friend had laughed at him and it doubtless wouldn't be the last.

He ignored them both as he slid his legs over the edge of the bed and stripped away the sticky pads and wires attached to the machines alongside him. Immediately, a furious beeping started up, and he slammed his hand against the buttons and dials. The ear-splitting racket got louder, and he gave serious thought to heaving the monitor over on its side, dragging it across to the window that Hone was staring out of, and hurling it through the glass.

Her phone in her left hand now, her dark eyes on the screen, Fang jumped from the end of the bed, reaching out to flip a switch, and the machine hummed and died. Fang was barely five feet tall, fifteen years of age, and a computing genius. North had met her in Newcastle when he'd been trying to track down a missing radio astronomer who'd befriended the schoolgirl prodigy. North had known the teenager less than six months and would trust her with his life. Occasionally, he felt guilty that he'd allowed Hone to recruit a teenager who still wore braces on her teeth. Mostly though, he simply felt grateful that she was willing to work alongside him. He accepted her judgement that he was a 'moron-person' as fair comment.

'You said that you remember one... two... two... six... zero... four. The only thing I can turn up in that sequence of numbers is associated with a colour, which is a very dark green. I've got some numbers of my own though,' she said, hopping back up. 'Tower "Bridge"...' – she emphasized the word 'bridge' and smiled beatifically as she did so, glancing up at him for a nano-second before going back to her phone – '... has a clearance of 8.6 meters closed, 42.5 meters open. You jumped—'

'The guy had knives.' North touched his fingers to the top of his head where the knife had cut him and they came away bloody. 'And I had a concussion, Fang. I didn't—' The hoodie he was pulling over his torso muffled his voice. The neck was too small for his head. He forced it through, feeling the stitches in the hoodie and his scalp rip apart one by one. '—jump. I took an executive decision.'

'Yeah, right,' said Plug. 'You're lucky I'm not burying you and your executive decision together.' He'd have done it too. Plug ran a successful undertaking business out of London's East End, which explained the black tailcoat, with a black silk top hat trailing silk ribbons on the windowsill next to him. Acting as support for a government agent was more of a side hustle for Plug. He enjoyed multitasking. So much so that the repatriation side of Padraig Donne & Sons had until recently involved the import of Class A drugs in coffins. He didn't do that any more. Not because of any scruples, but because his wife found out.

Hone had arrived with a holdall bag of clean clothes acquired from God knows where, slightly too small and slightly too big. He'd dumped it on the bedside cabinet without comment. Did he keep a random collection in his car for when agents got wet or shot or unexpectedly naked? Or maybe they were the clothes left behind by the agents who got wet or shot or naked. Suspicious suddenly, North peered at the fabric of the hoodie for bullet holes.

'Sidebar: a lad who lived round my way took an executive decision in Magaluf.' Fang aimed her remarks at Plug, as if North wasn't even in the room. 'His mam has to do everything for him now, bearing in mind he's a paraplegic. She tells people she doesn't mind—'

'—but she totally thinks he was a right twat,' Plug finished the story for her.

And, pulling on jeans that were too big around the waist, North glowered at them both.

Plug winked at him and the teenager laid the phone on her knee and started to re-plait the two stubby plaits that framed her round face. She kept talking. 'Anyway, the electro-hydraulic raising system of the bridge is computer controlled, so the engineer up in the control room couldn't override the lift at first. At 42.5 metres, the angle of the bascule is 86 degrees. The tide was high at 5.8 metres and moving at some speed, but—'

'Plug, lend me your shoes,' North said.

Leaning against the wall, his arms crossed, the six-foot-seven ex-cage-fighter looked down first at his own highly polished black leather shoes and then over at North's mismatched trainers, one too big and one too small.

'Not happening, mate,' he said with conviction.

This was unfair, North thought. No one was going to be looking at Plug's feet when they could stare at that Neanderthal brow and jaw, the scars and lumps and zigzag nose smeared across the red skin and shorn head, the gappy teeth and the complete absence of neck. Those who loved him saw only the kindness in his eyes. Those who didn't said there'd never been a man born uglier than Plug, but they said it quietly and some distance from the cauliflower ears.

'—you dropped into a high tide,' Fang said, still focused on the topic of North's near-death experience, 'and you managed to do it feet first with your arms above your head, which, as it happens, is the most survivable way.'

'I knew that,' North lied.

Fang glanced across at Plug only long enough to roll her eyes. 'The upper survivable limit is actually 56.69 metres or 186 feet. Even so, I estimate going into the water from the height you did is equivalent to the impact of a London bus hitting you while travelling at 29 miles per hour.' Now she sounded resentful, rather than appalled, that she hadn't been able to make the jump with him and get hit by the same theoretical bus.

North and Fang had first worked for Hone under duress, but united by a common enemy. An extra-governmental agency called the Board, which had once employed North as an assassin. When North realized those who made up the Board had ordered the death of the woman he loved, he'd wiped them off the face of the earth. But North was too good at what he did for Hone to let him go. And where North went, Fang followed.

For the best part of the past three months, they'd been working within the Friends of Cyclops. It was Hone's fiefdom. He reported to the Director General of MI5 and was given leeway other operatives couldn't dream of. And if he wasn't given it, he took it. National security and justice trumped the rule of law in Hone's judgement. For North, working as a government agent gave him purpose and when your life could be snatched away from you if you sneezed wrong, purpose mattered. He didn't know how long he had left to live. Not many people got to live at all when they carried a bullet around in their brain. But, however long he had left, he was making every second of every minute of every hour of every day matter. North didn't want to die. He wanted to live to a ripe old age. But if he was going to die sooner rather than later, then it would be doing his duty by his country.

'Shame you didn't have a jet suit,' Plug said. 'There's a jet engine pack at your back and two on each arm. And you get to fly like Ironman. Or a motorized hoverboard, they totally rock. Or a hoverboard motorbike – how cool are they?'

Fang tapped the keys of her phone, pressed on what North guessed was YouTube footage, and he heard a roar of engines. 'Yep, every superhero should be able to fly.' She looked up, her face apparently clear of malice. 'You'd never have to set foot on a bridge again.'

'I like bridges, Fang,' North said with infinite patience. 'I like the fact they take you from "here" to "there". I'm planning to carry on using them.'

The one-eyed man cleared his throat. He'd finished on the phone and turned to watch without any of them noticing. 'If you lot are quite done,' he said, extracting a packet of cigarettes and a box of matches from the pocket of his waxed riding coat.

His glance took in Plug. 'Some of what I'm about to say is classified. When you walk out of this door, you forget what you heard. Understood, Mr Donne?'

Plug adopted a look of innocence that sat strangely on his ugly face. 'Fang wanted me here, so I'm here. But I've a memory like a sieve, guv'nor.' Plug hadn't used the term guv'nor since he and North had been locked up together. North knew it carried not one ounce of respect. It was more of a warning.

'Our assailant's name was Jonathan Gaffney. Generally called Jonny.' Hone settled himself in the room's only chair – a straight-backed vinyl number in livid orange – and balanced the cigarettes and matches on the arm of it. 'Forty years old, born and bred in Edinburgh. Used to work as an accountant, first in Glasgow, then in Edinburgh. No criminal record and no history of violence. Now, there's three dead and five hurt from the attack, but thanks to you, he had to keep moving and the general consensus is that it could have been worse. The interesting thing is that Gaffney's been living on a remote Scottish island called Murdo, at the far reaches of the Outer Hebrides.'

Fang let out a small mew. Hone looked at her, but she gestured for him to carry on as she started tapping the keyboard of her tablet. Hone sighed. 'As I was saying, he's been living up there as a member of a pseudo-religious community known as the Narrow Yett.'

'Yett?'

'It's another word for a latticed iron gate – the kind you'd have in a castle but that's hinged on one side.'

'It's a cult?'

'They call themselves a cyber church. But yes – a cult by any other name. We've had a watch on them for over a year now. It came to our attention that Gaffney had left the island and travelled down to London. As you know, we placed him under loose surveillance.' Hone raised an eyebrow. 'Too loose, as it turns out. We'd planned to bring him in and see what was going on up there. Shame we didn't do it sooner.

'We haven't had time for a full post-mortem yet, but according to the pathologist at the scene, after you jumped, Gaffney crossed his arms with his knives still in his hands and damn near severed his own neck. As for the numbers you remembered—'

'Earlier, you mentioned five... eight... two... five... seven.' Fang's eyes were on her phone screen. 'I know what that's about.'

The rank smell of Gaffney's breath and sweat filled North's nose again. 'What?'

'Partial coordinates. Latitude coordinates for Murdo are fifty-eight degrees, two minutes and fifty-seven seconds north.' He'd been right. The man had said the word 'north'. 'The only problem,' she said, her face screwed up in concentration, 'is that pattern one... two... two... six... zero... four isn't the longitude.' She moved the tip of a plait into her mouth and started sucking on it.

The one-eyed man extracted a cigarette from the packet as he spoke. 'According to our theologians—' It was news to North that the security service had its own theologians, but these days it made sense. '—in his mania, it's likely Gaffney believed that his actions were morally significant. There's a certain ritual element to it all – getting to a high place, sacrifice, self-aggrandisement, et cetera.'

The one-eyed man struck a match to light his cigarette and inhaled as it caught. 'And three guesses...' – smoke came out along with the words – '... what Jonny Boy was missing.'

Fang spat out the tip of her plait. 'A head,' she said.

North felt Gaffney's hand all over again, saw the glinting knife high in the air, the realization he had only seconds to escape the descending blade and that the only way out was the void. He shuddered, and wished he hadn't, as his body splintered in pain and nausea washed over and through each and every bit of him.

'Testicles.' The cigarette bounced up and down on Hone's lower lip in time with the word.

Plug winced.

North remembered the leather bag that had swung from the cord around Gaffney's scrawny neck. Hone's one eye was on him, watching him work it out. North raised the tips of his fingers and brushed the skin at his throat and the one-eyed man nodded. He grimaced.

'That is beyond sick.' Fang missed nothing, North had time to think. 'He could have been wearing them as a talisman.'

Hone shrugged. 'It's possible. According to his ex-wife, Gaffney got caught up in the Narrow Yett when he went over there to advise them on a tax issue. The next thing she knew, he'd quit his job and cleared out every penny from their joint accounts. He even sold the family home in Morningside without her knowing anything about it. He walked out on her and their two sons and refused all subsequent contact. In his absence, she divorced him and up until a few days ago had heard not one peep from him. Then he called her.'

'What did he say?' Fang's black eyes were rapt. 'Was it: "You remember those testicles I used to have…"?'

Hone shook his head as if his patience was wearing thin. 'The Narrow Yett believe reality is in fact a computer simulation. Gaffney told her that he wanted the family to be ready to escape the simulation through "the yett". That it was about to open and going through it was the only way to save herself and the boys. He said she was to keep the boys out of school on Thursday, watch the Narrow Yett's YouTube channel, and that everything would become clear. Then he hung up.'

'That's it.' Fang slapped the palm of her hand against her forehead. 'The guy was an accountant. Numbers were important to him. One… two… two… six… zero… four. That's noon on Thursday, the 26th of April. The day after tomorrow. Actually, I suppose it's Wednesday already, which makes it tomorrow.'

On any other face, Hone's expression might have passed for a smile. 'That gives us thirty-six hours. Our analysts believe the cult's leader, Enoch Fraser, is planning to announce his own jihad – his own holy war. They've pored over his broadcasts and writings. They think he's going to "reveal" that this "simulation" has come to an end and humanity, as we understand it, faces annihilation. Because of this, he will order his followers to take the lives of those around them. Anyone dying at the exact moment the gate opens automatically rewrites their code. With a rewritten code, they can pass through the yett and go on to the next simulation. He's justifying murder.'

North thought about Gaffney's words: 'You'll be first in line, Brother.' Gaffney thought killing North would put his victim front of the queue when the gate between this simulation and the next opened. He thought he was doing him a favour. North shuddered.

'The guy was a nut,' Plug said. 'Who cares what he says?'

'His two million followers care. If we don't stop him, we could face hundreds – maybe thousands – of Gaffneys rampaging through our streets with guns and knives. That can't be allowed to happen. Which is why I want you up there, North. He can't make that broadcast. If you have to take out Enoch with extreme violence, do it. I won't have more innocent people dead because of this man's lunacy.'

Fang sat up straighter. 'Are you kidding?' she said. 'Gaffney killed people. Send in the army if this is an island of violent whack-jobs.'

Hone's face was impassive. 'Enoch is the problem here—'

'"Enoch?"' Fang said. 'So you're on first-name terms?'

'Prince. Elvis. Madonna. Enoch. His followers on that island aren't violent whack-jobs as far as we know. They're ordinary people living in a community with a distinct belief system. But if it turns out that they're all like Gaffney, North will be in a position to warn us. We've gamed the options. At the minute, we believe the best course is to send a sole operative. Someone who can operate independently, and who can improvise if they have to. Ideally, North, you take care of business, and you pull out before they even realize what's going on. We don't want carnage on Murdo either. To state the obvious, North, we want it quick and clean.'

Fang wasn't dropping it. 'But why does it have to be North? You only ever want us involved if the job is dirty and dangerous, and you think your agent could end up getting dead.' Her brows were drawn together in a ferocious scowl. Fang hadn't forgiven Hone for threatening her mother with deportation last year. She wasn't the forgiving kind. North knew her loyalties were to him and not to their boss. Never to their boss. Her code was her own and she didn't need Hone telling her what she should or shouldn't think – should or shouldn't do.

Plug stood straighter, as if ready to flatten Hone if Fang gave the word. If she was worried, then he was worried. North, though, felt a pulse of excitement. Hone couldn't walk in a straight line. He already guessed he wasn't telling him all of it, because he never did. But with Gaffney on the bridge, he'd started something and now he wanted to finish it. Fang seemed to sense what he was thinking.

'I have two words for you, moron-person,' she said. She counted the words off on her thumb and index finger. '"Testicles" and "pouch". If you go to Murdo, you could come back speaking in a really high voice.'

'I'm invested, Fang,' North said, moving towards the door. 'That guy on the bridge could have killed me. Damn near did. Hone's right – I can't have more innocent people dying when I could have done something to stop it.'

'Not so fast, young man,' boomed a voice from the door.

4

'You get to my age and you become invisible.' Dr Linklater glared at her patient. 'Did you know that? Men just don't see you.' Her tone was sharper than one of her surgical scalpels. Behind her, he sensed several junior doctors take a mutually agreed step back into the shadows.

A less invisible woman was hard to imagine. Professor Caroline Linklater had the appearance of a standing stone, and one that was currently blocking the doorway and scowling. 'Judging by the fact you're wearing a coat, nor do they listen. I told you to stay in bed and rest. I should have been at home hours ago. If you do a disappearing act on me again, North, I will wash my hands of you.'

North winced. He'd hoped he could get out of the National Hospital for Neurology and Neurosurgery without Linklater being involved. He'd first come across her when he was invalided home, having sustained a traumatic head injury in Afghanistan. She was brought in to consult by army medics. She kept him on her books when he quit the army, but after he joined the extra-governmental agency the Board, they'd put him under the care of a Harley Street doctor instead – someone who asked no questions but was generous with the psycho-cognitive drugs. Apparently, Linklater bore a grudge.

The consultant's eyes narrowed as she took in the fact that

the machine monitoring him had been disabled. Fang sat a little straighter on the bed, but North didn't think the *Who, me?* routine was fooling anyone.

Linklater's laser-like gaze moved on to Plug and then the writhing and squirrelling smoke in the air around Hone, who sat with hands miraculously empty of any offending cigarette. He gave her a slight nod. 'Caro.'

'Edmund.'

North hadn't realized they knew each other. Were they old friends? If so, the consultant didn't seem pleased to see him. 'I've told you before, those things will kill you.' She barked out a theatrical cough, then snapped the manila folder in her hand back and forth in front of her face. 'Am I keeping you from something, young man?' Her tone was icy and North sat back down on the bed. He didn't fancy his chances in any fight with Linklater – she'd fight dirty.

'Your MRI results are not what we'd want for a man with a bullet in his head.' Linklater must have missed the day they covered 'bedside manner' during her years of medical training. 'They show inflammation around the casing – probably due to the impact from the bridge, or again, when you landed in the water. Either way, the next few days are critical when it comes to intercranial pressure. So, if you're planning another disappearing act, don't. It may settle down of its own accord, but if you deteriorate...' – North felt his heart lurch in his chest – '... it will be fast and we will need to go in.'

North kept his face blank, but clenched his fists in frustration. Recovery from a brain operation would take months and he wasn't going to end up in a better place if they took a bone saw to his skull.

Linklater handed the file to a subordinate. 'In the meantime, we'll treat with mannitol and intravenous neurosteroids and monitor you. Your one and only job is to rest up. Understood? No more Action Man stunts for a while at least – and certainly no more jumping off bridges.'

'He didn't jump – he made an executive decision.' Fang's voice was mulish. 'He's a hero, not an idiot.'

Linklater spared her a glance as if to say the jury was out – before turning back to North. 'Not many veterans who sustain the kind of traumatic brain injury you did live to tell the tale. I take my duty to all my patients seriously, but I consider it an honour to treat veterans who've put themselves in harm's way. And, since you are back under my care again, I want to run a full evaluation and get the psych eval people in here. Any aggression?'

Did she mean like killing people? Because how many was too many? They were all of them really bad guys.

'Depression?'

He'd be happier if Fang didn't call him a moron-person quite so often.

'Insomnia?'

Sleep was overrated. He figured you could sleep when you were dead.

North shook his head. 'No, to all of the above.' He pasted a well-rested, peaceable smile on his face.

Linklater sucked her teeth, head on one side, as if she knew he was lying and it didn't surprise her. 'You bailed on me last time, North. Don't do it again. If you allow yourself to be exploited by ruthless people who don't care about you, it might end up costing you everything.'

He said nothing.

She tried again, her tone softer. 'You can't save everyone, North. You don't have to be a brain surgeon to know that. Occasionally, you have to save yourself. You owe yourself and your friends here that much.'

Linklater stalked to the doorway, before swivelling on her heel. She cast a final contemptuous look at Hone. 'And Edmund, I don't count you among his friends. If this young man chooses to disregard what I say because of you, I will hold you responsible if he dies out there. Which means I will be forced to hunt you down, and let me make it perfectly clear – it will not be pretty.'

And she swept away, followed by her juniors and a nurse. They'd barely cleared the door when Hone stood up. The smouldering cigarette end emerged from the cavern of his hand. He flicked the butt out through the barely open window and it dropped onto the sill, where it balanced half off and half on the stonework.

'I've a car at the door,' he said. 'I'll drive you home. Pack a bag and we can have you up on the west coast of Scotland by eight.'

'I'll go,' Plug said, his arms folded over his barrel chest.

Hone looked him up and down and gave a mirthless laugh. 'You're only in this room because I accept that working with North requires a certain flexibility on my part.'

The look on Plug's face would have quelled a riot. His voice was low and terrifyingly reasonable. 'Fang rang me. And North's my mate. Whatever they need, I'm down for. Do you have a problem with that?'

'Let's get this straight, Mr Donne. This isn't improv night at the local comedy club. Aside from your lack of official standing within the security services, your lack of experience in the field or indeed training, and, of course, your significant criminal activities in the not-so-recent past, you are the least undercover person I've ever come across.' Plug squared his Neanderthal jaw. 'Other than that, of course, you're perfect.'

'It's all right, Plug. I'm good,' North said.

Fang gripped the hem of North's coat and tugged him back as he stepped forward. 'Bozo, this is serious. If you go, you could drop down dead.' She wasn't giving up. 'Hone, his head might explode. Didn't you hear the doctor? He's not fit for the job.'

'He's either in or out, Fang,' Hone said. 'He can either hack this life or he can spend whatever time he has left doing Feck-Do-I-Not-Give.' The Belfast man's tone was ice cold. As if he didn't care. And it wasn't an act. Hone had bigger things to care about – like the safety of the nation state. A couple of months ago, Hone was willing to kill a British general and have the Home Secretary wrongfully committed to a psychiatric clinic for the risk both men posed to peace. He was willing to destroy the world's first conscious artificial intelligence, because he thought it too posed a risk to the future of humanity. He trusted his own judgement and nothing and no one else in this world. But the one-eyed man didn't sail under a false flag when it came to North's recruitment.

The one-eyed man had spelled it out – the fact North had a bullet in his brain was one of his most attractive qualities, because it meant he was disposable. And Hone saw great virtue in the idea of North being disposable. Because it meant that he could ask North to do whatsoever he wanted him to do, regardless of risk. However long North had left would be about doing his duty, because that was what gave his life meaning and Hone knew it. That was their deal. The more dangerous the job, the more likely it was that Hone would use North, which made him wonder what made this job on Murdo in particular so dangerous. 'In or out, North?' Hone said.

The thought of Linklater or anyone else cutting him open appalled North. He'd rather die on the job than on the operating table, a million times over. Plug knew this, because he would be the same. But Fang's huge brown eyes behind her Joe 90 glasses were on him; the light from the strip lighting overhead wiped out her expression, but her small fingers held tight to his coat tail. Could he do this to her? She was a genius, but she was still a kid and she needed him. Even if she wouldn't admit it. If it came to it, he could always refuse an operation, but she was right – he should stay and get the IV steroids at least.

'Send someone else,' Fang said, and now there was a note of pleading to her voice, which he'd never heard before and which he knew she would hate herself for. She was tired, he thought. And North felt ashamed that his situation was turning her inside out. She shouldn't be in this life. It was selfish of him and of Hone to exploit Fang's skills the way they did. They knew from experience that the world was full of bad people and terrible choices and they had that much in common. That they'd bought into this life of shadows without qualm, prepared to do whatever it took to stop the worst from happening.

But Fang should be in university, at MIT or Imperial, studying something that would make the world a better place, not standing alongside him doing battle with monsters. She was better than that and he wanted more for her than that.

And, of course, Fang was right. He should listen to Linklater on this one. It would only be a couple of days. This was a job like any other job. He had no compunction about taking out Enoch if he was head of some death cult that wanted to see the streets running with blood. He was extraordinarily good at what he did, but the simple truth was Michael North wasn't irreplaceable and Hone could easily send another agent, or think again about sending in the New Army. It was a wacko cult up in Scotland, not a terrorist organization in the Levant with a nuclear bomb. When he first took a bullet in his brain, he was fully prepared to die at any second, the risks he took irrelevant next to the certainty that the bullet in his brain would kill him sooner rather than later. He'd believed love wasn't for the likes of him. He was alone and that was fine. Now, he had a family in Fang, Granny Po and Plug. He was a government agent with purpose. Thanks to Fang, he had money. And thanks to a woman called Honor, he even knew what it was to love. And that the pain that came with losing that love wasn't enough to kill you. That you could still survive and thrive and cherish every second you had left.

North would put his own oxygen mask on first for a change.

'One more thing...' Hone said, as if he could sense the shift in North's thinking. 'There's a woman. She's been out on the island for five months making some reality TV show about the Narrow Yett. She's called Mia Barton and she was in a children's home the same time you were. Our records indicate you may well know her. We think she's our way into this. She'll trust you because she knows you, and time is of the essence. You can work on her. Get her to open up about how worried we need to be.'

North's eyes locked with Plug's, and he saw the same astonishment there as he felt. How many hours had he spent talking about Mia when they were locked up together? Too many. How many years was it since he'd heard the name?

Mia Barton had been twelve and he thirteen when they met. She'd be twenty-seven now. He'd thought of her a thousand times since, and thought about finding her a thousand times more. To explain, to apologize, and to beg for forgiveness that he'd let her down when she needed him. He'd tried to reach out at first. Calling her from the payphone in the young offender institution. But the carers had refused to put her on the line. They'd never told her he called. He knew it. He was a bad influence and they wanted him out of her life. Eventually they stopped accepting his calls. And then when he joined the army, he'd thought about it again. He was respectable. He'd served his time and he was remaking his life. But his nerve would fail him. After all, Mia Barton had moved on. What did she need Michael North for any more? He was a creature from her past and Mia deserved a glittering future. He'd considered it again when he started work for the Board and when he had money. But his existence then was one of blood and violence. He refused to risk infecting her with his darkness.

But if she was stuck on an island full of crazy, that changed things, and it all began to make sense. Because he hadn't understood when Hone said to watch Jonathan Gaffney tonight. Surveillance wasn't his go-to and MI5 had their own teams of watchers. But when he'd asked Hone why, Hone said he could do with the practice. North had thought it odd, but no more than that.

North's relief at still being alive blew away like dust.

His happiness of a few minutes ago – irrelevant.

He watched Hone watching him. Hone was always going to send him to the island, not least because North already had a possible source – Mia Barton. And North was always going to agree because he'd just got a taste of what was up there – madness and death, and living among the likes of Gaffney was nowhere he wanted to picture Mia. He couldn't allow Mia to get hurt. And if he didn't go, that's exactly what might happen.

He was frightened. Not for himself. But for her.

North headed for the door. He heard Fang say his name again and it hurt him, but he didn't apologize, he couldn't explain, and he didn't look back.

5

Lucy Sampson had kept a diary since she was young. These days, though, she had no choice in the matter. Writing things down was the only way she remembered them. And she didn't think the shock had helped her memory problems. It was still hard to believe he was dead. Her twin brother was dead.

'Laurie.' She said his name aloud and felt a hot tear escape her eye, but she fought against the misery. Grief, she knew, wiped her memory like a damp cloth against a blackboard at the end of a school day. Did they have blackboards any more? She supposed not. It was strange how well she could remember her schooldays but couldn't remember what she had for breakfast or lunch. Toast? An egg? Had she even eaten? She scrunched up her face, grasping for the memory, but it slipped from her. She could check the diary she kept up as she went through the day, but it wasn't important enough. Not like this meeting. The papers trembled in one hand. A Waitrose bag for life bulging with more papers was clutched in the other.

'Miss Sampson,' the voice called out in the huge marble reception.

Lucy didn't move.

The man who approached her was as thin as the blade of a knife. He was forty-something, in a tight black suit that, in her opinion, must have made sitting impossible and his left shoe squeaked as he walked. It was a brown leather brogue, she noticed, which he wore without socks. She considered that strange. Such bony ankles. Hardly a thing of beauty. Thought about writing it down, but didn't think there would be time.

His cadaverous face with its high knobbly cheekbones wore a tight smile, as if to say he was doing her a favour by being here at all. 'Miss Sampson? Lucy? It's very late for you. Shouldn't you be at home?' The accent was nasal and American.

He was talking to her, she realized. He had to be, because he was looking straight at her, and although she could not have told you her name right this second, the name he was using had a vague familiarity to it.

'Yet here we are again, Miss Sampson. I'm going to ask you to leave now, and these nice gentlemen will help you find your way out.' Lucy felt huge hands take hold of her. There were sniggers from the security men, who she hadn't noticed before.

They lifted her off her feet, so that only the tips of her shoes touched the marble, as if she were nothing and no one. But she used to be someone, she thought. She was almost sure of it. And she became aware of how ridiculous she must look. A middle-aged woman carried along by these huge oafs. Complete overkill. It wasn't as if she was dangerous. The thin man walked alongside and she sensed his mind was elsewhere. That he had worries of his own. That he found her an irritation. And something about him squeaked. Squeak, squeak. Like a—? Like a—? The huge glass door swung open and the thin man stood then, watching the security men as they took her to the furthest edge of the pavement as if their instruction had been to carry her as far away as possible. As if she might otherwise contaminate the building. The papers and her bag for life still clutched in her hand.

Lucy's distress rose the further she got from the building, but she could not say why. Perhaps it was because it was night-time and the streets were dark, despite the lights on in the offices and the headlights of the passing cars. She turned and discerned a flicker in the thin man's eyes but could not decide if it was contempt or pity or fresh irritation as they dumped her unceremoniously on the ground. At least, she thought, this time she was still standing. Although, she could not think when there had been another time. One of the security men smoothed down her skirt, which had rucked up, and winked at her. 'Let's keep you decent, eh, love,' he said, not unsympathetically.

The thin man scowled at him and then at her. 'Miss Sampson, you will be served with a cease-and-desist letter. Another cease-and-desist letter, I should say. You can't come back here, Lucy, do you understand?'

Lucy opened her mouth, then closed it again. What was happening? Why was she here? In the middle of the night? Panic taking hold of her, she clutched the plastic bag closer to her bosom. Where indeed was she?

The sign by the door of the huge glass building read Benson Corporation, and she had a sense that she had been here before. But where was she? Who was she? Trying to remember what was important. Trying to remember anything at all.

She looked down again at the papers clutched in her hand. The top page was in handwriting she presumed to be her own. *They still won't see me*, it read. *But Laurie is dead. Remember that, and come back. You need to know what happened to him. You must ask them what happened to Laurie. Put the papers back in the bag. Don't lose any of them. Ask about Laurie.* She did as instructed, and noticed the thin man was already walking away. Squeak. Squeak.

'What happened to Laurie?' she called after him, but her voice was drowned out by a passing red night bus and he didn't turn around. The glass door closed behind him.

Mouse, she thought. His shoe squeaked like a mouse. Through the huge plate glass windows, she watched the thin man on the phone. Mouse spread his hands out as if he was explaining something, his stance deferential. He's talking, she thought, about me. Turning his head, the thin man's cold eyes met hers for a brief second. He frowned as if she were a problem he was required to solve, and she felt goosebumps rise up along her arms. *Someone walked over your grave*, her mother would say, and hug her. And she missed her mother. Missed Laurie.

It took an hour for Lucy to make her way home. The key she wore around her neck helped, and the phone she had chained to the inside of her handbag linked to the Apple watch she wore kept her right when it came to catching the right bus and getting off at the right stop. Avoiding the night-time revellers and drunks. Carefully, she checked her watch and noted the day and the date on top of the piece of paper to indicate her last attempt to talk to someone at Benson Corporation. Her eyes ran over the dates. There were eight of them. Eight attempts to talk to someone senior about how her brother died, and why.

6

Hone dropped North home, where his agent changed into clothes that actually fitted him: the T-shirt and hoodie, each with the compass pointing to true north, that he'd picked up in Camden market, black jeans, a black goose-down jacket, his old army boots and a rucksack. Plug had made his own way home and Fang's face was turned to the glass throughout the journey, as if the Londoners still walking the streets were a source of fascination to her. Back in the car, en route to the helipad, Hone provided comms, a new Sig Sauer P226 and a wallet bulging with cash and platinum credit cards. North was no longer an ex-squaddie and government agent. He was the Eton-educated only son of an East End millionaire property dealer. Daddy was called Andrew. And Mummy was Fiona. Apparently, North himself had drifted around the world and never settled to anything. He'd been addicted to drugs and was still addicted to alcohol. He'd even done six months on a possession charge. His father despaired of him.

'Will any of this stand up to scrutiny?'

Hone made a face as if to say he didn't need to ask. '"All warfare is based on deception",' Hone said. 'Sun Tzu.'

'You didn't answer my question.'

'It's all good.' Hone handed him a tablet. His name and photographs of him on holiday in sunshine destinations and on the ski slopes, surrounded by partying friends, peppered social media and the internet. It would have been a nice life, North thought. Shame he was such an entitled fuck-up.

'What about my accent?' North had adopted accents before. He could pass for a member of the monied classes if he had to.

'You're a posh boy who doesn't want to be a posh boy,' Hone said. 'Your own accent works perfectly.'

'And Mia? She knows who I am. She knows my background. What's to say she won't give me away?'

Hone shrugged. 'Mia Barton is a potential source of information for you, and for us. After all, she's been living among them for months – she should know if they pose a wider threat. And she's a journalist. If something's wrong up there, she's going to be all over it. If you can persuade her to talk to you, it will give you a head start, which will save you time and effort. She's supposed to be a shortcut for you. A solution. Don't make her a problem.'

'Is Mia the reason you want me up there rather than anyone else?' He sensed Fang listening. 'Because you should tell me if you think this is going to get messy. If that's really why you want me to go.'

Hone regarded him evenly through his one eye. 'Focus on your mission – shutting down the broadcast and neutralizing Enoch. If you fail, innocent people will die.'

North noted that Hone hadn't denied it could get messy. But if Murdo was a dangerous place, then it was even more important he was there to protect Mia.

'If Gaffney was following orders,' North said, 'these people are going to be expecting an official approach from the police. If one doesn't come, that's going to be suspicious in itself.'

'That's true. If he was following orders, they're going to be ready for the police to rock up and make enquiries. Which makes it even more important that you get yourself onto that island as quickly as you can. Of course, there's always a chance Gaffney was a lone wolf. The best we can do is to put a security advisory notice on Gaffney's name getting out into the media – no one will publish it, for now at least. We've provided guidance that the perpetrator had right-wing extremist links and was under surveillance. There's a chance the name will come out online of course, but hopefully we can delay that till we know the broadcast isn't happening and you're off the island.' Hone ran his index finger along the purple scar that carved through his socket and into his cheek, as if it hurt and he was trying to soothe it.

North had no idea how Hone had lost the eye. The only explanation Hone had ever provided indicated that whoever had taken it paid for it with his life. There was never any doubt that Hone was a ruthless and exacting bastard of the first order, but then so was North. Unlike North, however, Hone lied as easily as he breathed. 'For the greater good,' he might say if prompted. Whatever that meant.

'Before you go, check this out.' Fang held up her phone and turned the screen around to face him. 'There's a court case here about an old couple in Bermondsey. They signed over their house to the Narrow Yett on the understanding they could stay there rent free till they died. The day the cult served them with an eviction notice, they killed themselves. That was three years ago. Their daughter took them to court claiming her parents were defrauded, but she lost the case because her parents were judged to have been in sound mind when they signed the contract, and there was a get-out clause re eviction. It boiled down to the fact that if the Narrow Yett saw fit to sell the property and invest the money in their mission "in this reality or in any other", the parents had no protection.'

Hone appeared irritated at the interruption. 'Sounds about right.'

'Real sweethearts, you're going to be mixing with,' she said. 'Just remember to bring back everything you're going out there with.'

North read the 'eyes only' theological briefing on the Narrow Yett on the helicopter flight. It was dry reading, but at least it was short.

The Narrow Yett is a movement that has been around for just over ten years. These days, it attracts techies, conspiracy junkies and those who would have traditionally been identified as 'God squad' types. At first sight, it would appear to offer a strong eco belief system, spiced up with tech, sci-fi and New Age values, appealing to the young and the old. Believers talk of 'authentic spirituality' and being part of a 'psycho-social, psycho-cyber movement'. Followers are big on restricting their eco footprint and its marketing and message is slick and powerful – save yourself, save others, save the planet. It has a strong cyber presence. Their icon is a gate with a snake curled around it, where the hinges would otherwise be.

Why snakes? North wondered. And as if the researcher who authored the paper had heard him, the paper continued:

'Yett' is an old English or Scottish word for gate. The gates were defensive, the latticed ironwork probably constructed on site and impossible to dismantle. Most of them are to be found in Scotland, with some in the border counties.

Enoch's use of the word 'yett' seems to be a nod towards the cult's Scottish origins as well as to the idea of the gates of Heaven, while the snake's significance is understood to be part of the simulation's origin myth of Adam and Eve and the Garden of Eden. It was a snake that persuaded Eve to eat of the Tree of Knowledge. Traditional churches have taught that was an offence to God and entailed a punishment of exile from the Garden. The Narrow Yett, however, maintains 'knowledge is power' and the snake is in fact a message left behind by the creators of the simulation that humanity needs to realize this isn't reality. Snakes – which have the capacity to shed their skin – have traditionally also been symbolic of renewal, transformation, immortality and wisdom.

There was an addendum.

Regarding their good standing or otherwise, colleagues at the Inland Revenue confirm they have an active file open on the Narrow Yett regarding tax avoidance. The Charity Commission reject the Narrow Yett's claim to be a 'cyber church' and a non-profit body. The Church of England is understood to have rejected demands for inter-faith discussions. The Narrow Yett are on record saying they have an avowed mission to spread the word that this reality is 'so much code and silicon'.

It is to be noted that on a dozen occasions, former followers have accused the leadership of abusive behaviours including sexual coercion, physical beatings and – within the past year – reports of mutilations. They have alleged that members are encouraged to cut themselves off from their families. Despite numerous allegations of abuse, including physical and psychological mistreatment, nothing has ever come of them. The leadership of the Narrow Yett bring in high-powered lawyers and the critics are shut down – it is believed they are either intimidated or bought off.

There was a scribbled note at the bottom from Fang. She must have scanned the briefing while he was getting his gear in the apartment. She'd written five words. 'Remember, bozo – don't trust Hone.' He dropped the briefing note into the small shredder next to him. He and Hone had reached an understanding the last time they'd worked together. Hone could ask him to do 'any crazy thing' and he wouldn't feel bad if it all went wrong. North hadn't demanded that Hone play fair with him. But, somehow, he'd expected it if he was working as a full-time member of his team. If Hone wasn't straight dealing, that put a different complexion on things. North had to be able to trust him. The problem was no longer whether he should or shouldn't trust Hone, the problem was it was too damn late not to.

7

Harris

The heady smell of peat and smoke rising from the Laphroaig whisky in front of North was enough to make his eyes water. The black coffee and the bowl of breakfast porridge, he thought, was little more than an excuse to drink the whisky.

He wondered how the barman would react if he asked whether there was a Japanese single malt among the 150 bottles of Scotland's finest the place boasted of. Not well, he was guessing. And he was almost tempted to do it, just for the hell of it.

'Excuse me.' The accent was a Yorkshire one and the voice like a silver bell. 'Do you have the time?' Through the mirror behind the serried ranks of whiskies, a pair of huge blue eyes met his.

North turned on his bar stool and checked his watch.

'Just past seven,' he said.

The girl's hair cascaded over her shoulders in blue waves the exact shade of her eyes. The kind a mermaid should have, he thought. 'Old school, huh? Most people use a phone. I like it. Thanks.'

She gave him the sweetest smile he'd ever seen, transforming the already pretty face into something breathtaking. Before switching her attention back to a friend whose own cosmetic rebellion involved a shorn head, a series of earrings along the cartilage of both ears, and a tattooed swag of green roses running along her forehead, dropping down her right cheek and onto her neck. Both girls had a tattoo of a snake curled around a gate on the inside of their wrists. Behind the girls, through the window, he saw that the promised rain had started to fall.

With a lazy but charming smile, he turned back around. He'd heard them order their own breakfasts and they looked settled enough, which suited him.

His bar stool wobbled on the uneven planks of the floor. North shifted his weight this way and that, but it made no difference. From the glint of amusement on the barman's face as he glanced across at his customer, there was a chance he'd once spent a satisfying night sanding the nub of just the one leg from every seat in the place to guarantee the discomfort of his customers.

North kept his body still, not to give the publican the satisfaction of watching him try and fail to find some ease. He switched his attention back to the Scotch whisky in front of him. He could almost taste its honeyed fire as the alcohol hit the back of his throat and burned a channel down to his stomach, but the aim of the exercise was only to look like a tourist, he reminded himself. He needed to stay alert. The porridge might help but the whisky wouldn't, although it would help with the aches and pains racking his body courtesy of the fall from the bridge. But he'd need a barrel of it to drown out the guilt he felt at the upset on Fang's face as they drove away from the hospital. He spooned up more of the porridge. The army had taught him that you ate when you could, because you didn't know when you'd get another opportunity. The porridge was cold, with a smack of salt on his tongue. And he wasn't an idiot. He knew he was taking a risk ignoring Linklater's advice. But he had made his decision and would live – or die – with the consequences. He scraped out the bowl then swallowed down three steroids from a stash Hone had inveigled out of the furious consultant before they quit the hospital.

But if he was being honest with himself, he was feeling nervous as well as guilty. Not at the prospect of going undercover in a cult that castrated its members. Though now he considered the issue, Fang was right and he probably should be.

At the thought of that particular horror, his eyes went to the girls. They looked arty and modern and content in each other's company. They were a far cry from the wreck that had been Jonathan Gaffney.

No, he wasn't nervous about what he might face in terms of the cult, he was nervous at the idea of seeing Mia again. And there wasn't enough whisky and there weren't enough pills in the world to make him forget his grim childhood. Mia belonged in his past – she had been a bright shining light, one of the very few. He wanted to remember her that way. If she was ruined, or vulnerable – if her life had been a misery and she was a mess, he would blame himself. Rightly, because he'd let her down when she needed him. He shook his head. He'd deal with that when he had to. The more pressing concern was getting across to the island in the first place.

Raising the Laphroaig to his lips, and keeping it casual, he scoped out the crowd of noisy breakfasters in the corner by the fireside. Americans. Big men, the pumped-up kind, whose conversation revolved around how much each of them could bench-press. They were military or ex-military, he couldn't decide which – all five of them in dark clothes and one of them, the smallest of the bunch, wearing his aviator sunglasses indoors. North set the tumbler back down having barely wetted his lips, but licked them anyway, savouring the amber taste of the malt.

At least memorizing his cover had passed the time when he wasn't poring over the map. Murdo was the most remote and least well-known island in the Outer Hebrides. A smaller uninhabited island lay north-east of it, called Porr, and an even smaller one to the south west, called Freyja. Seven stone stacks surrounded it, each of them with their own Gaelic name. The archipelego looked a long way from anywhere. Even by chopper, it had been three and a half hours from London to the Isle of Harris and it was quarter past six by the time they landed at Stornoway airport. A senior police officer had met him off the flight and driven him the hour and a half to Auchterburgh where boats left for St Kilda and for Murdo. North sensed the policeman's curiosity but, to his credit, he'd filled the time with conversation about the imminent storm and sport.

The police officer advised the Salmon Inn was as decent a jumping-off point as any if the idea was to get across to Murdo. North had planned to establish his credentials as a lone traveller looking for adventure, to turn the conversation with locals to island-hopping, and then find a boatman willing to make the journey across to the island. But with the two girls already camped out in the bar, he was changing the plan.

He ran over his mission in his head. Stop the Narrow Yett broadcast due in twenty-eight hours. Then take out Enoch – with extreme violence if necessary. Stopping the call to arms should be possible if he destroyed the equipment. If Enoch got hurt in the process, North wouldn't feel bad. Those who lived by the sword and all that. Hopefully, without their leader to turn to, Narrow Yett members would realize that violence was not the way to go.

He checked the earpiece was still in place. The line between him and Fang in London should be open by now. Perhaps she'd managed to grab a couple of hours' sleep? Or was she making him wait on purpose? Punishing him because he was here instead of being tucked up in hospital being poked and prodded by Linklater. But he wasn't going to apologize and he wasn't going to explain. He didn't see why he should have to, and he wasn't sure if he could. He knew Fang had been lonely before she teamed up with him. That classmates punished her for her genius and difference by bullying and excluding her. That she'd never had so much as one school friend. But Fang had a mother who loved her and a grandmother who would breathe for her if she could. What did she know of hungry, loveless childhoods where violence and fear were the norm? A childhood like that changed you for ever and not for the better. Mia's friendship, though, had given him faith in himself at a time when it would have been easier to give up.

He'd let Mia down once and still felt guilty about it. If there was a prospect that she needed help now, he needed to be there for her. Plus, this was his job. He'd signed up for trouble and that was okay with him. Better than okay, it was good. Trouble was what he was made for. It wasn't up to Fang to keep him from harm. The kid was a genius, but not even a genius could keep him away from Mia.

There was a crackle down the line and Fang started talking, her voice distant and tinny. He caught the words '...*Come in, moron-person...*'.

Fang and Granny Po had set themselves up in a luxurious suite in the Empire Hotel overlooking Hyde Park. Fang could have bought anywhere she wanted in London, courtesy of the money she'd stolen from the agency he used to work for. But she said Granny Po appreciated the convenience of room service and enjoyed teaching the housekeeping staff how to clean more efficiently. Plus, she said, the Empire had the fastest Wi-Fi speed of any hotel in London. The suite was certainly a far cry from the flat over the Oriental Dragon takeaway in Newcastle's West End where she'd grown up and where her mother still lived. That had been squeezed between a boarded-up betting shop and a burnt-out laundrette. And it was nowhere that was ever going to contain the phenomenon of Fangfang Yu.

As for the stolen money, Fang had already given away millions, only keeping enough for 'basic needs'. North figured the teenager was still a millionaire many times over. But thanks to her hacking genius, he was just as wealthy. He didn't take it for granted. No one who'd grown up in poverty ever did. And he felt no guilt about the theft. The Board's original purpose had long since been corrupted by the powerful and its obscene wealth sorely misused. Fang was an infinitely better custodian.

He imagined her working at the polished table in the dining room she'd set up as her office, the tips of the massive lime, plane and sweet chestnut trees visible through the window over the parapet. The last time he'd been, Granny Po had hung her washing on a line she had rigged up on the terrace and old-lady smalls snapped and flapped in the breeze, drying high above some of London's most expensive real estate. He could only imagine what the hotel staff made of the tiny old woman and the stroppy teenager.

Reception wasn't great. He leaned his ear on the palm of his hand in an attempt to ram the earpiece in further. He was counting on the fact that, from a distance, he looked exhausted or hung-over, or both. Still wasn't good. Frustrated, he contemplated pulling out the Iridium 9575 Extreme satellite phone stashed alongside his gun in the rucksack leaning against his stool. It was GPS enabled with a one-touch SOS button, but anyone watching – and North always assumed someone was watching – would wonder why he didn't have a standard iPhone or android. He concentrated harder instead, zoning out the ruckus of other people's conversation and laughter, the chink of cutlery against crockery, the dull commentary of TV sport set above the bar.

It was enough. *'...this cult leader Enoch was born in Ayrshire. He's forty-one years old. He spent a couple of years in a seminary before he was thrown out on the grounds "his temperament was considered unsuitable for the priesthood". No idea what that means.'*

North noted that she hadn't asked how he was feeling. Which was fair enough. He'd made his decision. Fang didn't want him arriving on Murdo without knowing who he was going to be dealing with out there. The kid might be young, but she was a professional.

'Enoch took a degree in IT at Glasgow University. Then temped as a software programmer in various tech firms. Even set up his own gaming company at one point. That's how he met Maria Maclean – she had an interest in start-ups. Liked to mentor up-and-coming tech talent. She was ten years older than him. And apparently she took the mentoring way too seriously because they were married within three months of first meeting.' Fang snorted. *'In an interview at the time, Enoch said it was a "coup de foudre" – love at first sight.'* There was a retching noise on the other end – Fang was not one of life's romantics. *'Her company provided financial software for processing payments and salaries that was being picked up by all kinds of big companies and governments. On their first anniversary, the happy couple went skiing in Italy with friends. A helicopter dropped them off-piste and, an hour in, she fell and hit her head. She died in his arms. He said he had to leave*

her body up there or he'd have died as well. It took two days to find her remains. All very boohoo.'

North shook his head, then realized the barman was frowning at him as if wondering whether his customer was insane. North waved a cheery hand. He was fine. He turned himself so that neither barman nor fellow customer could see him talking.

'Bit of a result for our main man though because he inherited the company. Three of the directors were related to the dead wife. Her mother and father, and a sister. It cost them 4.5 million to buy him out, which all but broke the company. Even so, the father was on record at the time saying it was worth every penny to see the back of him.

'According to a magazine article, Enoch says he then spiralled into drink, drugs and a sex addiction, till he woke up one morning naked and covered in vomit to find a white shining light spilling out of a thin high gate in the middle of his penthouse bedroom. It opened and the Big "E" understood everything in that moment. That reality wasn't what it seemed. That it was a computer simulation devised by a civilization with immense computing capacity. And it's a civilization that isn't just running Planet Earth right now, but one that has multiple simulations on the go – many of them way better than this one. That Enoch was being given the chance to escape this simulation with all its failings into a different and better simulation. He obviously thinks he got that offer because he's so special. Instead of passing through this gate though, he decided to stay and to warn others that nothing about this life is real, that they needed to accept that fact, and be ready to leave the simulation when the "yett"– opened again.

'The chosen one quit the coke and easy sex, called himself a cyber prophet, and spent the next four weeks writing the

Simulation Gospel of Enoch. I've skimmed it and there's a smattering of genuine physics in there along with a great deal of pop science, New Age bollocks, eco-warrior stuff, Bible verses, Norse mythology, mixed into a puke-inducing stew of any conspiracy you care to name...'

'Like what?'

'Usual stuff. Lizards run the world. 5G is designed to give us dementia. Dogmen built the pyramids. Democrats eat babies. Vaccines are bad. Governments give you cancer. The usual twattery, much of which he ties back to the simulation argument, errors in the coding, glitches, evidence, et cetera.'

'So he's talking rubbish?'

Fang made a *hmmm* kind of noise, which meant she was thinking so fast her words couldn't keep up with her thoughts. She settled on: *'Yes, when you talk lizards. No when you talk about us living in a computer simulation. The chance we are doing just that is put at around 50-50 these days by plenty of astrophysicists and important thinkers. Elon Musk...'* – she was hitting her stride – *'... puts the chance we're living in base reality at "one in billions".'*

North decided to drop the subject, at least for the moment. If this was virtual reality and his experience designed primarily to be experienced by others, he preferred not to know. It would also, he thought, explain a lot. And if he kept thinking about it, it would explode his brain, which had enough problems as it was.

'And how did Enoch end up on Murdo then? Shouldn't he be in LA or San Francisco?'

'The wife left it to him along with everything else. It was in her family – she'd been left it by her grandmother. He wouldn't give it back in the settlement, so the family let it go. Ten years ago, he set up base camp there with a bunch of unwashed groupies and it's gone gang-busters since.'

'Give me a minute,' he said to Fang. Judging by the loud voices and flushed faces, the men in the corner liked booze as much as coffee with their bacon and eggs. The remains of five breakfasts and a great many empty glasses littered their table. Were they perhaps climbers or divers over on holiday? The biggest of them, with the neck of a serious weightlifter, started braying at some dirty joke, and North caught the eye of the smaller man next to him, who sat very still with his back against the wall, the aviators now pushed up on to his head, his face a mass of ancient red-and-white scarring pulled so tight it left his startling white teeth exposed. North figured the scars for burns and him for the boss. If he had caught it correctly, the guy's name was Kincaid. His gaze still locked on North, Kincaid made a chopping gesture through the air with the side of his hand and the group quietened. North wondered if he had done it because his colleagues were becoming rowdy, or merely to prove he had that kind of control. North didn't want to be the one to look away first, but neither did he need the wrong kind of attention from some alpha male out to prove a point. He glanced away and hated the fact the guy would think he'd won.

Still, he was intrigued. He raised a finger to call over the surly barman.

'The guys by the hearth...' He kept his voice low. Hoped the barman wouldn't look across. 'Are they here climbing?' He pointed at the coffee maker and held out his empty mug to keep the barman busy and near him.

As he reached for the coffee pot, the barman made a face. At least, North thought that was what he was doing. It was difficult to tell between the greying whiskers and the long-ago etched lines of disgruntlement. The old man shook his head. 'Nope. They're contractors, working across on Murdo.' North sat to attention on his stool. 'Waiting for some buddies.' Were they connected with the Narrow Yett?

'What kind of contractors?'

'All I know is they work a month on and a month off. These guys were out on the lash in Glasgow all last week.'

'I thought there was nothing on that island,' North said.

'*Good for you playing dumb*,' Fang said in his ear. '*You're a natural.*'

The barman took North's money and shook his head. 'There used to be an old rocket tracking station on there. They've switched it out to a "spaceport".' He made some phlegm-filled noise of disparagement at the back of his throat. 'They're planning to send satellites up into space.'

He didn't give North any change as he moved off to the end of the bar.

North kept his voice low. 'Can you check it out, Fang? It's a small island. Maybe they'd be helpful.'

'*I wanted to follow the money. There's something odd—*'

The voices behind him went up a notch; he caught a gale of raucous laughter and glanced back at the girls. Orange juice and the remains of egg rolls were in front of each of them. They'd angled their bodies away from the men in the corner, as if perhaps they'd come across their type before and didn't care for them. Neither did he, although the contractors could be his ticket onto the island if he played it right.

'Just do it, Fang.'

North heard an obscenity and swung round to see the faces of the men at the corner table turning dark and ill-tempered, although Kincaid's own face sported a sardonic smile. The biggest man was walking away from the girls with a bottle of champagne in his meaty hand, his knuckles so white North was only surprised the bottle neck hadn't sheared clean off.

The champagne slopped over the rim of the bottle and onto the wooden floor as he walked. As he neared his mates, he attempted a laugh to show how little he cared, but the noise that came out was something thin and mean. North was guessing the big man had bought the bottle in an attempt to ingratiate himself with the girls, who weren't interested. Good for them. He smiled despite himself. The contractors were an odious crew. And in any event, he had figured out an alternative way onto the island. Why take the easy way?

'*"Just do it?" Quick reminder, bozo: you're not the boss of me.*'

North grimaced. She was mad at him. He should explain he had stuff going on here. That this was a sensitive play and he'd misspoken. Perhaps he would have, but she started talking again, her voice sulky. '*Where it really started to happen for Enoch was seven years ago when he moved his mission to the net. And it took off again after some Hollywood A-lister started tweeting about living in a computer simulation. The theory picked up a head of steam and websites and papers ran features with all the usual suspects quoted, but this time round the Narrow Yett kept getting namechecked. Partly because of Enoch's millions and tragic backstory. It gave him traction and respectability.*'

North cut her off again. More obscenities and complaints. The girls were judged a lost cause. 'Stuck-up dikes,' he heard as the big man collided with the mermaid girl's chair, almost knocking her from it, then carried on regardless. Enough already.

North stood up from the bar and blocked his path. 'I think an apology's in order, mate.'

8

The big man was swaying on his feet, reeking of last night's booze, but not what North would have described as incapacitated. Surprise and then the promise of violence flared behind the pink eyes almost lost in pouchy eyebags.

Behind him, the others could tell trouble was brewing. One of them called to their friend to leave it, but North didn't want him leaving it – he came in closer, stepping into the big man's personal space, his own shoulder knocking heavily against its opposite number. Primitive but effective as a call to arms. A signal that announced: *You may be big but I'm bigger and a whole lot more dangerous.*

North picked up where he left off. 'It's the twenty-first century, mate. Women don't need a Yank tosser buying them a pair of silk stockings and thinking they're owed. Me, on the other hand? Pour me a glass and I'm anyone's.' He winked.

He heard a laugh from behind him, but didn't know if it was from the mermaid or the girl with the rose tattoo.

Maybe it was the laughter? Or the word 'tosser'. Or maybe he considered North's wink disrespectful? Because it was certainly meant to be. But, grunting, the big man swivelled on his heel and readied himself to throw a punch. He telegraphed it from so far away that he might as well have been doing it in slow motion. North moved his body away and to the side while balling his fist and driving it into the other man's ribs. Satisfaction blossomed at the core of him. Whoever said violence was never a solution got it wrong.

From across the bar, there were shouts of protest and chairs scraped against the wooden floor as three of the other contractors came to their friend's aid. North risked a look at the girls. For his plan to work, they couldn't get dragged into this. But he relaxed as he realized they looked as if they wanted to get out of there as much as he wanted them to leave. Their coats were in their hands and they were halfway across the floor. The girl with the mermaid hair turned at the door. The blue eyes met his – full of worry and something he couldn't name. She gave him a tiny apologetic smile as if asking his permission and he gestured she should go, that he was good, just as her rose-covered friend tugged her away with a sharp 'Come on! We have to go!' He felt the hit of cold air as they pulled open the door and it swung shut behind them.

There was an excellent chance the girls considered him an idiot, he thought, as a punch came out of nowhere. And there was a good chance they were right, he thought, as it landed on his jaw. He was outnumbered, which was bad, but his back was to the bar, which was good, because he didn't want them surrounding him.

He had to play this right. If he messed this up, not only was he not getting a lift across to Murdo so that he could save Mia and the lives of thousands of innocent civilians, but he was going to get beaten to a pulp at the very moment he was supposed to be eating jelly in a hospital bed.

He let out a sigh. Dr Linklater had said rest. She hadn't specifically said *Don't get involved in a fist fight*, but she doubtless thought he had more sense. He leaned back to avoid another punch to his head, then did a rapid one-two on the guy who had thrown it. His hand stretched out along the bar, searching for his whisky. He found and then threw the glass with as much force as he could muster at the big guy, and then another and another. The glasses bounced off cheekbones and temples but only shattered when they hit the floor, the smell of wasted whisky saddening the air.

The bar was narrow, which was also good as it blocked the Americans' route to him, and unless they rushed him, he was all right.

They rushed him. He picked up a ladderback chair, swinging it high and wide, and it smashed against the tallest of the group, who staggered but didn't drop. Reaching for another, his grip was broken as it was kicked from his grasp, splintering as it flew across the bar. North overturned the table instead.

His own temper exploded as someone landed another punch and his body remembered the hurt from the fall into the Thames. Fight. Wound. Kill. His temper slid away from him in a blood mist, but he hauled it back and got control of himself. These guys were fighting as a pack because he'd challenged one of them. Even though he'd shown he could fight and that there was a price to pay, they were going to keep coming, which was a problem because North didn't want to miss the tide either.

He had miscalled it. It would have been simpler to have started a conversation, bought these knuckleheads a round of hard liquor and thumbed a lift across.

The barman was shouting. Something about the police, but North knew he would be long gone before the police arrived. But he did need to make it to the exit, because otherwise he'd miss his chance with the girls. He wrenched the old-fashioned phone from the bar, keeping his hand over the earpiece, and walloped it around his body to bring it with full force into the side of the head of the main offender. The big man went down as if he was never getting up again. North shifted the weight of the phone to do the same thing into the face of the next guy and felt the nose crunch under it and then the larynx. The man's hands clutched at his throat as he fought for breath. Two down. Two upright. Kincaid, to North's surprise, was still seated and watching the show with apparent interest.

Pivoting, North raised his boot and smashed it into the knee of number three, who let out a scream as loud as any North had ever heard. Three down. One upright. The boss, now chewing on a toothpick, made no effort to help the wounded.

North's anger was gone entirely now. Instead, he felt a cold-steel focus and – he'd admit it to himself if no one else – a lick of pleasure. Like stretching after getting out of bed. He could do this all the livelong day – except he couldn't, because he had the tide to catch. This final guy had been at the rear of the group; maybe he moved slower, or maybe he was the least keen to tangle with him? Either way, that didn't mean he wasn't trained and capable. Even as North was considering his approach, the back door along from Kincaid swung open to reveal another two Americans with the same haircuts and tattoos. Under his breath, North swore. The faces of the new arrivals were confused as they took in the bodies of their three buddies on the floor. The one still standing.

'Kincaid?' one of the newcomers called over, as if waiting for the word to move.

What none of them knew was that North had a knife sheathed in his boot. He resisted the temptation to take the fight in a whole different direction. Instead, he held up both palms. 'Hold on, guys. Your buddy swung the first punch.' North addressed his remark to the newbies. An outright lie, but from their line of sight, they weren't to know. Their buddy didn't seem the temperate sort, so it didn't sound impossible.

'I don't want trouble.' He addressed his last comment to Kincaid.

'Bit late for that, Limey,' Kincaid said, moving the toothpick to the corner of his mouth, and his voice was scratchy and soft and dangerous.

North picked up his rucksack and slung it onto one shoulder as he backed up to the door. This was all in the timing. If he messed up the timing, this whole thing was a bust. As his gaze flicked between the three contractors and their seated leader, he twisted the brass handle. The snick caught. He twisted again. They were moving forward. Twisted a bit further. They came on faster. The fight was over, but it didn't look like they had got the memo.

He eased his way around the door and pulled it closed behind him just as he heard the Americans hurl themselves against the wood. He hauled it shut, grabbing a dog lead from a hook in the foyer.

He knotted it with a constrictor knot, then looped it back onto the cast-iron hooks draped with Barbour waterproof jackets. It wouldn't hold for long and there was always the back door.

The familiar taste of blood in his mouth, and North started to run.

9

New York

The call between Benson's global head of security, Brett 'the Mouse' Munson, in London and Benson Corporation's chief operating officer, Theodore 'Teddy' Benson, in New York didn't go well. For a start, the Mouse had to wake him, which Teddy always hated.

It took Teddy a cold minute to clear his head enough to absorb what the Mouse was telling him. That some madman who lived on Murdo, Scotland, in that cult – did Teddy remember there was a cult on Murdo, Scotland? Did he remember everything about the cult on Murdo? Yes, Teddy remembered.

Good, well the lunatic had lost it on Tower Bridge, London, and gone on a murderous rampage. Mouse had it all on the best authority. From Benson's friends in high places over there.

Teddy stood up and left the bedroom quietly, grabbing his black and gold Versace dressing gown as he went. He didn't want to wake the escort whose company he had enjoyed last night.

In the living room of his penthouse, he shrugged himself into the robe then poured himself a large Patrón tequila and downed it in one, still listening to the Mouse. Gaffney had killed three people and injured five more.

Teddy swore. He grabbed the remote and turned on CNN,

keeping it on mute. Tower Bridge was an iconic landmark, he figured they'd get round to it. Courtesy of her arthritis, his mother didn't sleep well and often watched TV during the night. He wondered if she was awake. She certainly wasn't going to be happy when she heard there was a connection with Benson, and she didn't know the half of it.

Teddy had fought in three wars. He'd been decorated for bravery and had killed too many enemy combatants to number. He was familiar with death, and all those men and women who had served with him considered him a brilliant strategist. In business, similarly, he inspired confidence, and in his private life he numbered friends who would have laid down their lives and reputations for him. But he sought the good opinion of only one person in this world – his mother, Mrs Elaine Jacqueline Benson.

His was a life of service – to his mother, to the corporation and to his country, and he was fine with that. She had his future all planned out and he wanted what she wanted. He had every confidence that it would work out exactly as she planned. A backstory of abandonment and adoption, because Mrs Benson shared no DNA with Teddy. The army. Medals. A period of years on the board in which he would be seen to make money, his own money, not just to inherit it. A Senate seat, and then a run at the presidency. Mother would pick him out a suitable wife – a blonde paediatrics consultant played well, according to their market research. The search was already on for this perfect wife, and when he met her he would be happy. Or at the very least, he would be President.

Mouse kept talking.

The fruit loop had then killed himself.

At least there was that, Teddy said. Mother liked things discreet, and you couldn't get much more discreet than dead.

The guy's name was Jonathan Gaffney, although the Brits were trying to keep it out of the press for as long as they could. The name would come out though, because these things always came out.

Teddy said nothing. But instinctively, his eyes went to the corridor that led to the master bedroom. He'd been reckless. He kept being reckless and it had to stop. He'd have his people come by in the next hour or two with an NDA for the girl. Or other people come by with a car that had a spade in the boot and a full tank of gas. It could go either way. NDA, he told himself. The girl was sixteen. Too young to die. Then again – the girl was sixteen. Too young to sleep with, even if he was only thirty-seven. If it came out, it would look bad for him. As if he preferred them young and fresh out the box. Which he did, but no one needed to know that. He wasn't a pervert – it was a predilection. So with regret, it would have to be the car drive to a remote spot and a grave in the woods. He'd make it right with the agency. He felt a pang of regret. He'd wake her for one final fling though. She'd want that if she knew what the day ahead held for her.

Shame he couldn't trust the girl, because she'd been cute as buttons. But the only person Teddy trusted was Mother and the guys he had served with on active duty. These were the men he employed for jobs like this. He paid them well and they would never betray him, because they believed in him. He was destined for greatness, and they were going to help make that happen. Mouse was one of them. He had saved Mouse's life. Without regard for his own safety, he had dragged him out of a building that had just exploded. Visited him in hospital, and when he was invalided out of the army, found him a job at Benson. He found all the men and women who'd served under him jobs, wherever they wanted to work. Invested in their businesses, or put them through courses if they preferred to retrain or skill up. And he made sure his charitable foundation spent a significant proportion of its annual revenues on veteran causes. He was a good guy who had to make tough calls, he reassured himself. Not that he ever doubted it. He was special. He was chosen.

He zoned back into the phone call as he poured out another shot of tequila. He didn't want to call Mother at this hour but when he did, he already knew how it would go. He imagined her face as she heard the news – the way her lips tightened, the lines around her eyes, the way the air around her grew colder. It was up to him to protect her from this kind of irritant. But seriously, Murdo was a pain in the arse. First everything that could go wrong up there had done. Then the satellite engineer Sampson getting cold feet, and now this. It was never easy.

Teddy finished the tequila and threw first the tumbler and then the Lalique bottle into the hearth with as much force as he could muster. Splinters of glass exploded over the ash and brickwork.

'Boss? Boss?' Mouse's voice was a tinny thing. 'This Gaffney guy has been on the island for years. Do you understand what I'm saying?'

For some reason, Teddy thought of the story of the Snow Queen and the boy she had kidnapped with the sliver of ice in his heart. What the hell made him think of that?

'What do you think we should do?' he asked Mouse.

He could hear Mouse sucking on his teeth as if he wanted to make sure they were secure and that no one could steal them.

'The Brits will have to send someone out to the island to investigate,' Mouse said. 'I've briefed the guys on the base. The Brits can't be allowed to find anything. If we think they're getting too close, we deal with it. Publicly though, we make nice and cooperate. Nothing to see here. We're a commercial operation, planning to put communications satellites up in space. If we're asked, the neighbours are odd bods and our guys are told to avoid them. Maybe once it settles down, we give it a few months and we put an offer in on the place? That way, the rest of them scatter to the four corners, which could only be a good thing, and we get the run of the island. Till then, there's just one problem.'

Teddy gritted his teeth so hard his jaw hurt. Was it too much to want solutions, not problems? 'What problem is that?'

'Not so much a problem, as a loose end.'

'I don't like loose ends.'

'The sister,' Mouse said. 'We don't know what he told her. She's another kind of fruit loop. Won't leave it alone – last night she was back here again. She doesn't know whether it's night or day, what's up and what's down. But bearing in mind everything that's happened, you wouldn't want her going to the press right now.'

Teddy's gaze went back to the window, to the glittering city and out to the world that was his for the taking. Up to the skies, which would be his too very soon. He knew who Mouse meant.

'That country is full of crazies,' he said. 'Too full, wouldn't you say?'

'We'd be doing everyone a favour, Boss. Including her.'

Both men were quiet for a beat. It was going to be a long day, Teddy thought. He didn't think he'd go back to bed. It would just delay the inevitable. 'And Mouse. Send someone round to the penthouse in an hour. I have a package that needs taking care of.'

10

Fang's voice was in North's ear again.

'*Let me guess, you had a fight.*'

'I admit there's a chance I could have handled it better,' he panted. 'Let the record show I'm disappointed in myself. Okay?' It was pouring, he realized. Raining so hard, deep rivulets sheeted the road.

'*You realize that one day you're going to meet someone who fights dirtier and better than you.*'

She was right because she was always right about everything. But that day wasn't today. He heard a crash and shouting behind him as the men in the bar forced open the door.

The quayside was further away than he'd realized. He could see the sea ahead of him now, piles of lobster pots, a steel container. The horizon. Mountains. A vast sky. He was horribly exposed. No place to hide.

He pushed himself to move faster.

The captain had already unmoored the fishing boat. Sodden green ropes lay in a heap on the quayside. The captain turned back into the boat and took hold of the wheel.

North started yelling, 'Wait for me!'

Behind him, the three men were closing the gap.

Both girls had turned towards the noise, but the captain was oblivious. His hand on the throttle.

Obscenities from the Americans coming thick and fast. One of them hurled a stone, which only just missed him. Another. And another.

The captain pushed the throttle forward and the boat moved away from the quay.

North readied himself. After all, it went so well last time he attempted a manoeuvre like this.

On the boat, the mermaid girl had placed her hand on the captain's arm as if pleading with him, and he was turning his head, but the boat was still moving in the wrong direction.

North picked up speed and felt the reluctance in his body to make the leap.

Three.

The nervousness that there would be a madman on the other side ready with knives, and no way out but down and disaster.

Two.

That the laws of physics would rule against him.

One.

He leaped high and long – his arms reaching, circling. His legs still running through the open air. Bringing both legs round to make the landing. If the boat hadn't started reversing, he'd never have made it, he thought afterwards. But then again, if the boat hadn't started reversing, he wouldn't have smacked himself against the wall of the wheelhouse so hard he thought it might have snapped his sternum in two. He'd have sworn but he had no breath left in his body to do it.

His pursuers were calling out from the quayside but, exhilarated and spreadeagled on the deck, he watched as the captain raised a cheery hand in goodbye then raised his middle finger as he pushed the throttle forward and the boat roared away – the shouts of protest all but drowned out by the engine. A fine sea spray landed like rain on his skin.

He managed a wheezy 'Morning' as the old man glanced down at him. Whiskers. A jaunty black cap.

He was in this now. If his head exploded, he'd be miles from any help. He was on his way to an island to stop the leader of a cult declaring war on the world. No one was standing in his way.

He must have been grinning.

'Aye well, don't look so happy, lad. It'll be a rough old crossing to Murdo. If you'd the sense you were born with, you'd swim back to your pals there.'

11

London

With North safely on the boat out to Murdo, Fang figured she could make headway on her own enquiries. She'd get to the contractors as soon as she had a minute. But with the clock ticking down the way it was on the cult, a bunch of American hardmen weren't the immediate priority.

What she had been trying to explain to North, before he picked a fight with whomsoever, was that there was something about the information on the Narrow Yett's website that bothered her. They certainly made great play of their good works, not just in the UK but globally. In fact, the vast majority of their charitable efforts were in the developing world. She searched and cross-checked the various charities and foundations listed. An orphanage in Tanzania appeared to have the same kids as a school in Uganda. The photographs were from the same shoot. Was it an attempt to keep marketing costs down or something else? Did they both exist? Did either of them exist? She pulled up a listing of a health project in Colombia. The Narrow Yett was channelling £200,000 into the development and running of a new health clinic in a remote part of the country. It was an ambitious and worthy project. Fang approximated an email address for the *Los Angeles Times* and drew up an email sign-off. She wrote as a researcher on the health desk. Would the new director of the health clinic, Mr Luis Moreno, agree to an interview with a journalist on the paper's health desk? In particular, the paper was interested in the progress of the clinic's construction, sources of funding, and the role of churches and charities in health initiatives across South America. She pinged it across.

North had ignored her pleas to listen to the consultant and stay in hospital, while Hone dismissed her opinion as if it was worthless. And the hurt she'd felt got worse after North told her to 'just do it'. She wasn't a child to be bossed around. Okay, technically she was a child. But she was smarter than both men put together, and now she was trusting her instincts, even if they wouldn't. Fang kept pulling up reports and websites till she found what she wanted – a charitable project in the UK. This one in London.

It was time for a real-life site visit, she thought. She didn't want North having all the fun.

And in any case, Plug looked way too comfortable watching breakfast TV on the sofa next to Granny Po. He was midway through his second room-service full English – plus French fries – and his third pot of tea when she suggested an outing to Bayswater.

Fang and Plug made their third drive-by. They were in the hearse.

'Tell me again what you think says "undercover" about a hearse,' Fang said.

Plug glanced across at her. 'We have a job on and the other cars are all booked out. My wife said it was this or the horse-drawn carriage. Believe me, there's nothing undercover about a pair of black horses wearing ostrich plumes and the word "Mum" spelled out in chrysanthemums.'

'This is not a discreet car.'

Plug rolled his eyes at her. 'People die all the time. This is a very discreet car. Plus, you'll have noticed that it comes free with a babysitter.'

It was Fang's turn to roll her eyes, but she wasn't old enough to drive yet, so she would reluctantly admit that Plug had his uses. Not least the fact that agreeing to Plug as an escort was the only way Granny Po would stay home. Granny Po had loved Plug from the start. Something about his ogre-like slab of a face and obvious capacity for violence enthralled the old lady. She did not feel the same warmth towards North. Granny Po believed North would only ever drag Fang into trouble but had complete confidence Plug would only ever drag her out of it. Granny Po was a product of her times. As a modern woman, Fang knew she could get herself into and out of trouble without help from either of them.

Plug glanced across at her. 'You're not happy about the fact he left the hospital, are you?'

Fang considered the question. 'Ever seen one of those old-time movies where some innocent sap gets thrown into a volcano to satisfy the gods who live at the bottom of it?'

'Nope,' Plug said. 'But now I really want to.'

'North's the idiot who jumps in, before anyone even gets the chance to push him.'

'Yet here you are, helping him,' Plug said.

Fang looked at Plug. The huge, scarred hands gripping the wheel. His craggy side profile with its louring brow and jaw. The broken nose. Plug didn't run from a fight either. 'Gaffney was a eunuch. Plus, he cut off his own head. If I don't fish North out of this particular volcano, he's toast.'

Plug didn't reply. Instead, he peered through his side window. 'Where is this place then? The GPS says we're here.'

Fang consulted her phone then pointed at a graffiti-covered building that had the look of a former community centre. The words 'Tomorrow's Citizens' Trust' were just about discernible on a tattered board. She checked the building next to it in case she'd made a mistake, but it was a second-hand car dealer.

She scrolled through the figures. According to the NarrowYett's official charitable spending, this place had been given £100,000 last year, and just under that the year before. It provided care and activities for 200 refugee kids in need. She held up her tablet with the photographs of teenagers playing outdoor ping-pong and gathered round a campfire in woodland. 'The blurb says it's based on a city farm.'

Plug pointed at a mangy cat rooting round the bins. 'You're milking it,' he said.

Fang climbed out of the hearse with a heavy sigh.

'If anyone asks, we're knackers and we're picking up a dead cow.' Plug locked up the car and started to whistle 'Old MacDonald had a Farm'. He stopped when the screaming from inside the building started.

12

North Atlantic

'You have to turn round.' The rose-tattooed girl was insistent. 'We can't take him with us. We're operating a strict No Visitor rule right now.'

That wasn't surprising, North thought. If Enoch was planning his very own holy war, he wouldn't want strangers in his camp asking why.

The captain gave her an even stare. 'Last time I checked, this was my boat.'

'But we're paying you. Take him back.'

The captain ignored her. Kept his eyes on the horizon. 'He's aboard now. I'm not losing any more time.'

Rose-tattoo girl looked furious, the mermaid bemused, the blue hair lifting in the early morning wind. She knelt down next to the unexpected passenger, regardless of her jeans on the wet deck. The smell of warm vanilla washed over him. 'This is a long trip, Handsome. You know that, right?' As she spoke, she did her best to tie back her hair with a scrunchie – the wind almost whipping it out of her hand. Close up, her smile was even lovelier. 'You're going to be stuck on this boat for two and a half hours till we get to where we're going and then you'll be on it just as long coming back. We don't want you getting hurt by those bully boys, but you're sure about this?'

He stood, reluctantly, one hand on the side to help himself up, then reached down his hand to her. She took it as she stood. Her hand warm and small in his. For a second, he fought back nausea. Yes, every bone in his body had remembered they could make his life a misery if they all worked together. The pain must have shown on his face. But she misread it as disappointment that he wouldn't be getting to follow her onto the island.

North tried to channel how a public-school wastrel of a rich boy would feel right now. He stood straighter. As if he was up for the adventure and didn't much care who knew it. 'Who doesn't love an outing and some sea air? It's the most perfect day for it.'

'And thank you for sticking up for us in the bar. We shouldn't have abandoned you, but Mac texted he was ready, so we couldn't hang around.'

'You shouldn't have interfered,' her friend piped up. Close up she was older, all razor-sharp cheekbones, tattooed roses and Liverpool attitude. She draped an arm over the younger girl's shoulder. 'You made things worse. We know those guys. We didn't need you going all macho and upping the ante.'

She had a point, of course.

North shrugged as he found a seat on a couple of wooden crates along the starboard side down from the wheelhouse. 'They were getting on my wick.' That much at least was true. 'If it hadn't been about you, it would have been about something else.'

Dimpling, the mermaid girl giggled. 'Did you really hit that big lad who bumped me? Because they wanted to put your lights out back there.'

'Well, I don't think they wanted to buy me a pint.' North winked at her. He could tell from the captain's stance that although he had his back to his passengers, he was taking it all in.

Rose-tattoo girl frowned but appeared to have given up the fight. 'Seriously? You don't even know where we're going.' Even so, she was fuming, he thought. He figured he had time to bring her round but the harder it looked like he was trying, the more suspicious she would get. He shrugged, settling back on the crate, his arms along the sides, turning his head away from the girls and out to the sea. He was going to have to play this cool. Eton, he thought to himself. I went to Eton and I think I'm God's gift. 'I swear you won't even know I'm here, ladies.' He tapped the side of his nose with his forefinger. 'Ignore me.'

He heard the captain gather phlegm at the back of his throat, lean to one side and spit with vigour off the side of the boat and into the sea. North didn't blame him.

The girls went below after a while. He heard raised voices, catching a loud '…just can't help yourself. Enoch specifically said no…' Rose-tattoo girl really didn't want him on this boat. Which was unfortunate, for her at least, because he wasn't getting off it. Did she know what Enoch was planning? Is that why she was so dead set against his presence? Or was she just compliant with the rules – whatever they were?

For a long while, the seaman at the helm said nothing and North let the silence sit between them until it took on an easiness it didn't have before.

Eventually, in a wheelhouse that looked as if it was held together by rust, the captain unscrewed a tartan thermos and poured himself a steaming black coffee. North raised his head at the scent of it.

'Not to push my luck, but you don't have any more of that, do you?' It smelled a whole lot better than the coffee at breakfast.

Grumbling into his whiskers, the captain found an old tin mug and filled it. He held it out and North stood to take it with a grateful nod. The sea under the boat moved as if it were alive and his body swayed with it, attempting to find a natural balance. The spray hit his face and he felt the salt already settling on his skin. He didn't know if he had done the right thing or the wrong thing flouting Linklater's advice, but being out at sea beat being in hospital by a long way. Not to mention the thought of seeing Mia again after all these years. He felt his spirits lift.

'Bearing in mind this is my boat, I'm thinking you should introduce yourself,' the captain said, his voice breaking into North's reverie.

He held out his hand. 'North.'

'Mackenzie,' the captain said, ignoring the outstretched hand, his eyes going to the wheel and then the sea again as if to telegraph that he wasn't letting go of the boat just for North. 'You can call me Mac.'

'Mac, I'm sorry to land myself on you. I was in a fix back there.'

The old man harrumphed, but North thought he was softening. 'Aye well. Better to be fleet of foot than lose the war, sure enough,' he said. The captain's eyes were on the horizon, one calloused hand on the wheel, the other gripping his own coffee. In profile, the old man's wild white eyebrows grew upwards to merge and tangle with his hair. 'The surprise is Comfort and Joy...' Mac sounded reflective.

'Comfort and Joy?' North said. Mac had to be kidding.

'Aye, the sweetheart is Comfort and Rosie O'Grady is called Joy. This time they're over on a supply run for booze – they've some kind of party happening. But often these two are across meeting their "guests" that have come up from London and

Edinburgh and the likes. Well-to-do types, generally speaking. And when those people head back all prayerful and reflective, they're heading back with lighter pockets right enough.' He took a close hard look at North. 'D'you ken what I'm saying, lad?' He raised the bushy eyebrows again and looked straight at him.

'I hear you,' North said, and his gaze went to the horizon and the darkening clouds.

The captain gave him a sardonic grin. 'Don't be trusting any of the folk on the island. They're a strange lot. And those that aren't strange, well, there are hard men among them. Harder than those Yanks. The kind you wouldn't have got away from.'

Mac's accent was soft and rolling and North had to strain to make out the words over the rush of the wind and the roil of the sea. The captain flicked another glance over at him. North said nothing. If they were usually on meet-and-greet duties, that would account for the fact both girls were such out-and-out stunners. They were part of the cult's public face.

The captain chuckled and shook his head – at the folly of young men, North thought. As if the sea dog had sensed North had stage-managed the whole thing to appear heroic and dashing, and was now embarrassed at the predicament in which he found himself. On a boat he couldn't get off with two girls he couldn't trust. The boat shifted under his feet and North staggered a step or two.

'Do you ever take anyone off the island who isn't in a good place?' he asked, keeping his voice low, and the old man didn't turn his head but under the wild brows his eyes flickered sideways to North and then back to the choppy sea. Was it North's imagination or was the sea getting rougher? He pulled his coat up around his ears.

'It's been known,' Mac said. 'I've heard stories right enough from those who've lived there months or years even, and they wake up one morning and can't wait to get clear of the place. Tales of snakes and beatings and goings-on I don't even want to understand. There's plenty of visitors leave content and prayerful, right enough. But then there's those they bleed dry who come off not knowing which way's up. Take my advice, son, when they invite you to stay – and they will invite you, wait and see. Comfort will bring Joy round, right enough. Pretty as a kitten or not, she'll have her own way. Joy may come across tough, but she won't be able to say no to her. Make your excuses, laddie, and I'll take you back with me.'

'You don't seem to like them much.'

The captain grimaced. 'They don't belong here. You can't own a place like Murdo – it owns you. Aye, these lot pay well enough for the back and forth. But none of us do it for the money. My family were real islanders. They were evacuated in 1929, a year before the folk on St Kilda. They were told they could be buried back on the island but the government gave backword on it. My mother was only a babe when she came off the island, but the family went back and forth till the war came. I've promised her when it's her time, I'll bury her on the island. This makes it possible. My boat's called *An Tilleadh Sàmhach – The Quiet Return*. Once that's done and my mother's at peace, I'm done with the lot of them.'

As he finished speaking, the boat hit the trough of a wave and dropped and then lifted then dropped again. North felt himself falling and grabbed for the wheelhouse as spray hit him harder in the face than he would have thought possible. His eyes stung with salt water.

'Hold on tight,' Mac warned. 'This is going to be a bad one.'

13

London

Plug waved Fang back to the car but she ignored him. He pushed open the door, which led into a shabby hallway. He held up his finger against his lips to Fang and picked his way through the litter and empty beer cans with surprising grace for a man of his bulk and shoe size.

Behind the swing doors ahead of them came a sound of more screams and loud weeping. Two heavy-set men with shaved heads and dark bomber jackets were dragging a wailing girl out of the arms of a skinny woman with cropped jet-black hair. The more the men pulled at the girl, the louder the skinny woman cursed them. Fang didn't know what language the woman was using. It was nothing she'd heard before, but she didn't have to speak the language to understand the fear and distress in the room. Around half a dozen more women were milling around the men – a couple pulling ineffectually on their jackets and sleeves, a couple more standing apart as if too frightened to get involved. One of the men let go of the girl with one hand and punched the skinny woman on the temple. As she reeled, she let go of the girl. With a yell of triumph, the men dragged the girl away. She screamed in frustration.

Nobody heard the swing doors open. Fang picked up the nearest thing to hand – a piece of rubble – and threw it full force at one of the men. She had never been any good at sport, but by some fluke, it smacked off his shoulder.

'What the—?' He turned, and as he did, he too let go of his grip on the struggling girl. The other man still had hold of her, but she'd seen her chance and kicked his knee. It was his turn to let out a shriek of pain as he went down. Free of her captors, the girl ran for the open door where Plug and Fang still stood. None of the other women followed. Instead, they backed away, huddling together against the furthest wall, their attention now on the apparently terrifying face and figure of Plug Donne.

'Any more of you fancy doing a runner?' Plug's eyes didn't leave the two men, but the remark was addressed to the women. 'How about you, sweetheart?' He glanced for the briefest of seconds to the woman with the cropped black hair, but she shook her head, then raised a hand and blew the girl a kiss.

'Okay then.' Plug now had a length of wood in his hand. 'Get this one out of here, Fang,' he said. 'I'll catch up with you when I'm done. Shouldn't be long.' Fang didn't want to leave him, but she saw the wisdom of what he was saying. Anyway, there were two bad guys versus one Plug. It was hardly fair on the bad guys.

She grabbed the girl's hand and ran back down the corridor and out through the exit, and down the street, turning right and left without any regard for where they were. Panting and exhausted, the two of them finally stopped running to cower in an alley that reeked of stale urine. Pressing themselves close to the brick wall, they sheltered behind bins piled high with rubbish. Fang raised a finger to her lips, and the girl nodded. She was pale, her face dirty and her hair matted and greasy. Fang didn't know how old she was. The same age as she was, maybe? Or perhaps a year or so older? She gave her what she hoped passed for a reassuring smile and the girl lowered her head, chewing on a knuckle, and Fang knew she was trying hard not to cry.

This was on her. The girl didn't know where she was. Fang found that she had crossed her fingers. How long should she give it before she rang Plug? Five minutes? Ten? She wouldn't want to distract him. Had she done the wrong thing by leaving him with the two men? What if one of them had a gun? Or a knife? Or they both did? What if Plug was bleeding out on the floor of that place? She'd never forgive herself. What was she going to say to his wife, to his kids, to Granny Po? Tears began to prickle her eyes. Perhaps she should leave the girl hidden here and go back to check on Plug.

There was the noise of an engine, and then the sound of a car reversing.

Fang pushed her companion behind her and peered through a gap between the bins.

The tail end of the vehicle reversed slowly and carefully past the entrance of the alley.

'Wotcha, girls!' Plug leaned out of his window and waved. 'Fancy a cup of tea and a sticky bun? I'm famished.'

Plug parked the hearse up close by Paddington station and left a note in it saying 'Picking up. Five minutes.' He knew a café close by. Fang had a Diet Coke, the girl hot chocolate, a sausage sandwich and a bowl of chips, which she devoured as if she hadn't seen food in weeks. Plug got his tea and another full English with chips. Fang held up three fingers and he ignored her. They sat at a table against the wall with Plug facing the doorway. Behind Plug was a jar on the counter full of what

Fang was beginning to think were jellied eels.

She dragged her eyes away from the jar and pointed at herself. 'Fang,' she said, then pointed at Plug. 'Bozo.'

Plug spluttered out a mouthful of half-chewed egg and bacon. Coughed. Pointed at himself. 'Plug,' he said. Pointed at Fang. 'Pain in the arse.'

The girl got it. 'Catina,' she said. 'I have English. Thank you for helping me. Thank you very much.' She went back to the chips. Fang met Plug's eyes. He lifted his hand and flattened it, stroking the air. She read it as advice to take it slow.

'You're welcome.' Fang kept her voice casual. 'How old are you, Catina?'

'Sixteen years old.' Catina put a hand in front of her mouth to cover up the fact she was eating. The sandwich was going down as fast as the chips.

Only a year older than her, Fang thought. 'And what was all that back there then?' she said. 'It didn't look good.'

Catina's eyes started to fill up again. 'I come for maid job in big hotel.' She used the heel of her hand to mop up the tears that had started to fall on the sandwich in her hand.

Plug made a face at Fang as if to say: *Now look at what you started.* He handed Catina a paper napkin and then another. His voice was soft, as if he was talking to one of his young sons. 'Let me guess, sweetheart. No maid job. No big hotel.'

Catina shook her head. 'I feel fool.' She put down the remains of her sandwich. Next to her, Fang could feel Catina start to tremble.

'You don't have to talk about it if you aren't ready,' Plug said.

But Catina stuck out her lower lip, her face all earnestness. 'I want talk.'

'Then we want listen,' Plug said. He sat back in his seat. It had to be serious, Fang knew, because there was still food left on his plate and that was unheard of.

Catina stared through the window at the passing traffic. 'We picked up at airport with Irini...'

'Was that woman trying to stop those men taking you away?' Fang remembered the skinny woman with the cropped black hair who had tried so hard to keep Catina from the two thugs.

Catina nodded. 'We travel on same plane from Romania. We talk. She know why we here. She do sex work at home. Nice girl. She told me "Run, Catina" at airport. But too late.'

Plug looked as if he was considering going back to the community centre and breaking whatever bones in the bodies of the two men he hadn't already broken. 'I guess it was Irini who said to tell you good luck. I tried to persuade them to leave, but none of them would come.'

Catina's expression darkened. 'I wish they all come with you. Most of all, Irini. I don't want them punished because I run away.'

Plug patted her hand. It had to be a real possibility.

'How long were you at that place, Catina?' Fang had once been in a bad place with a bad man herself. North had saved her then. And she had shot the guy, which had made her feel altogether better about the situation. But it had been a narrow escape.

'Three days. I tried to – how you say...?'

'Escape?' Plug interjected.

She nodded up and down with as much emphasis as she could. 'Those men they catch me. They say they teach me lesson. The other girls – even ones who think maid job – they have family back home. Irini – she has a seven-year-old daughter, who stay with her grandmother. Men say family hurt if we escape. Irini would not want her daughter hurt. These are bad men. But I have no one back home. My mother – she die two months ago.'

'Why were they keeping you at that place?'

Catina shrugged. 'Men say they wait for more girls come on next plane and then we all go…' She pointed upwards. Her tongue found the word. 'Scotland.' Another reach. 'Glasinggow.'

'Glasgow?'

Catina nodded. 'Yes. Glasgow. Work for important man.'

Fang had rung Hone in the car as they'd driven to the café. By the time the police arrived, the women at the centre and the two injured men had gone. Fang hoped at least some of the women decided to take their chances.

Catina froze at the sight of the police car pulling up outside the café. As the officers got out of the car, they were met by an older woman in sensible shoes and a ponytail. She looked both businesslike and exhausted. Fang was guessing she was a social worker. The three of them took a moment to talk to each other before they stepped into the café.

'I want stay.' Catina looked nervous, suddenly, speaking fast as if she knew her time with Fang and Plug was running out. 'Do you think they will let me? I get job. I promise I work hard.'

A gust of cold air pushed its way into the café as the social worker opened the door.

'I don't know,' Fang said. 'I hope so.'

'Catina?' The social worker raised her hand in greeting, offering the three of them a tired smile. 'My name is Sheila. I'm here to take you somewhere safe. Would you like a minute to say goodbye to your friends?'

Catina stood up. She gave a small shrug, as if to say: *Why delay the inevitable?* Her dark eyes, when they met Fang's, were anxious. 'Don't let them send me back, please. They cannot hurt my family but they can hurt me.'

The social worker stood back to let Catina go by, signalling to the police officer to open the door. She didn't touch the young girl, her outstretched arm just hovering behind her, guiding her new charge through the door, out into the street, and into the system.

Fang and Plug watched as first Catina climbed into the back of the police car, and then the social worker.

'What do you think?' Fang said.

'No chance. They'll send her back.'

Plug wasn't saying anything Fang hadn't already thought.

'She's a witness though,' Fang said. Outside, the car pulled away. Fang couldn't see Catina any more, only the social worker's profile.

Plug went back to the congealing food still on his plate. 'I doubt she can tell them much. Certainly not about the operation over here. Maybe something at home, but that's probably a bad idea. Anyway. Are you thinking what I'm thinking?'

'That someone from the Narrow Yett was skimming or even diverting its allegedly charitable giving? This is bigger than that though. The cult seems to have links with organized crime.'

Plug nodded. 'Yes. And sex trafficking isn't for amateurs. Before those two guys at the centre took a nap—'

Fang adopted an inquisitive expression. He ignored her.

'—they volunteered the information that they worked for one Craig Drummond, a Glasgow gangland boss with fingers in all kinds of pies including drugs, intimidation and sex trafficking. ...'

Fang sucked up the last of her Coke. It was flat and warm

and sickly. 'And what's Drummond's connection with the Narrow Yett?'

'I have no idea,' Plug said. 'But I know the right man to ask, and his local is a stone's throw from here.'

14

North Atlantic

An hour out of Harris, a squall blew up, the boat bucking and rearing as the sea took it and flung it this way and that. There was a bang and a scream from down below and North figured one of the girls must have fallen. As he stood to check, a surge lifted the boat up into the air and then fell away and the boat dropped back down, knocking him off his feet.

The captain clung to the wheel. 'Tell them to hold on down there,' he said. He wrestled with the wheel and North staggered to his feet, the deck now at a forty-five-degree angle. He half walked and half crawled to the steps of the cabin. 'How are we doing?' he called down.

Comfort answered, her voice barely audible above the waves and the wind. 'Joy fell and hit her head.'

'Mac says to hold tight,' he said. 'And stay below.' He could make out the two girls in the darkness. They were clinging to each other, with a foot of water on the floor of the cabin. He wondered about shutting the doors but then again, if the boat went under the girls needed a way out.

He turned back to the old man at the wheel. He had a radio in his hand but, as North watched, the wind sheared the antennae clean off the roof of the wheelhouse. North put his finger to his ear to see if he was still in contact with Fang, but there was only silence. He went to try again and, as he did, a massive wave swept across the boat, knocking him from his feet and smashing him into the port side. The boat was sizeable and solid, but nothing could withstand this kind of battering for long.

He sensed a shift in the boat as it began to turn. What was the old man doing? If the boat turned side on to the waves, it was going to roll. North shook his head in a bid to clear the seawater from his stinging eyes. Mac was gone – the wheel on the move. The captain had been washed overboard.

He stared over the side of the boat, narrowing his eyes, trying to keep most of the spray and the rain out of his vision even as he started yelling for the girls.

Comfort emerged first, holding tight to the brass rail at the top of the stair. She looked pale and frightened, and he didn't blame her.

He pulled on the rope attached to a cleat and hoped to God the captain knew how to tie a knot better than he did. He pointed at the wheel. 'Grab that and hold on to it for dear life or we're all dead.' Water smashed against the deck as he looped the other end of the rope around his waist. This was possibly the stupidest thing he had ever done in a long list of extremely stupid things. 'Keep her pointing straight ahead into the waves. Don't let her turn or the sea will have us over. Go.'

Comfort was quick on the uptake. She was on her feet and across the deck faster than he'd thought possible.

Joy was still on the steps, gripping the brass rail. She was bleary, a red welt across her temple apparent even under the green of the roses. 'What are you going to do? How can I help?' she said as he stepped up to the side of the boat and paused for a second, scanning the horizon. A shadowy mass in the distant water.

He didn't answer.

Was that the old man?

He wouldn't get a second chance.

The boat dropped into the trough of a wave and North dived into the darkness.

The cold was like something from another world. He felt the weight of the rope haul him backwards and downwards. There was too much of it, he realized too late. It was coiled together and sinking fast. He started swimming. Long powerful strokes to get as far away from the boat as he could, moving all the while towards what he was hoping was Mac, putting distance between him and the ravel of rope, stretching it out. He felt the tension ease a little – the boat was going one way and he was going the other. How long did he have to reach the drowning captain before the rope hauled him back? A minute? His lungs felt like they were bursting as he fought his way to the surface and drew in a massive gasp of air. He turned his head this way and that, trying to glimpse the old man in the water.

The rope tugged at him. He could make it out snaking across the choppy sea. If the boat went down, he realized, unless he did something soon he was going down with it. And for a man with a bullet in his brain, drowning would be a strange way to go.

He caught a sudden glimpse of something a few yards away before it disappeared – the waves blocking his line of sight. He took a breath and dived below the surface, twisting and turning. There he was. The captain's oilskin had acted as a buoyancy aid, but it was beginning to take in water now. North reached for the old man's sleeve and took hold as he hauled him upwards. But as they broke the surface of the water, the rope bit into his midsection with sudden violence and he imagined the boat dropping into a trench of water then lifting up the other side. His grip loosened on the slippery sleeve as the boat itself hauled him away from the captain.

Frustration and exhaustion swept through him. He set his teeth. He was not letting Mac die out here on his own. He reached for the Gerber Ghostrike knife in its sheath in his boot. If he cut the rope, he was cutting his only chance of getting back to the boat and the captain might be dead already. No point them both dying. He took hold of the rope and sawed through it, tiny fragments of jute floating up through the water. His fingers already clumsy from the cold, he slammed the knife back in its sheath, or tried to. He felt it slip from his grip and imagined it floating down to the seabed.

But the knife had done the job it was meant to do. He had his freedom. He dived back into the water and behind him felt the rope disappear into the distance.

15

London

'No Teeth Francie' was an old friend from Plug's cage-fighting days. If by 'friend' you included those you once beat to an apparent pulp on a semi-regular basis. He was slumped in the corner of the Shamrock, under a picture of Pope John Paul I, his hand raised in blessing. There were only dregs left in the glass, and Plug didn't have to ask. Despite the fact it was not long after ten in the morning, 'No Teeth Francie' would expect a pint of Guinness. The barman was ahead of him. Plug paid for the two pints and made for Francie. At least he'd lined his stomach this morning.

'If it ithn't "The Fighting Irish". You've not got any better looking then,' Francie said, nodding as Plug put down the other man's pint. It wouldn't do to spill a drop.

Plug made sure to sit a little distance from his old sparring partner. The nickname No Teeth was something of a misnomer. The only teeth Francie lacked were his four front teeth, upper and lower, knocked out and swallowed in his very first cage fight. Spittle could be an issue. But since Plug had been the man to knock them out, complaining was not an option.

'It's hard to improve on perfection, No Teeth.' Plug put down his own pint. 'And I see the rumours aren't true then.

The word on the street was you'd gone over to Turkey and got yourself some fine implants.'

Francie held up a finger while he took an egg from his pocket, cracked it on the side of the glass, watched as it disappeared under the foam, and then took a long slow drink. There was the sound of his tongue smacking against the roof of his mouth. He dropped the eggshell in the ashtray on the bench next to him. Now Plug looked, it wasn't Francie's first egg of the day either.

'True enough. I did,' he said, pinching off the foamy moustache with his thumb and forefingers. 'But weren't they knocked out in a fight. Still ea'thy come. Ea'thy go.' He shrugged. 'And how'th the undertaking busine'th treating you?'

'They're still dying and I'm still digging, Francie.'

'Sure enough, we're never short of the dead.' The other man gave him an even look. 'What i'th it I can do you for, Plug? I'm not ready to be measured for one of your coffin'th. The Devil can wait for me a while yet.'

Plug glanced over to the bar and held up two fingers to the barman. Then leaned in towards Francie. 'Craig Drummond. You've spent time in Glasgow. Talk to me.'

Francie's mouth turned downwards, and he made a show of blessing himself. Then rummaged in his pocket with his other hand, pulled out a dental plate and slipped it over his upper gums, settling it with his thumb. 'A regular terror. You'd be best not asking questions even this far from him.'

Plug shrugged. 'Indulge me.'

Francie appeared to think about Plug's point, then nodded. 'All right.' He lowered his voice. 'Drummond is big time. Fingers in a lot of pies. Drugs, girls, intimidation. He runs the doors up there. Owns a couple of nightclubs, half a dozen bars. Even owns a casino.'

'Tell me about the girls.'

'Usual story. Ships them in from some godforsaken spot. Gets them hooked on drugs if they weren't before. Puts them out to earn. When they're done, brings in another load.'

Plug looked down at the new pints that the barman had slid in front of them. And thought about the woman who had tried to help Catina get away. He should have insisted she come with them. But it was too late for her now.

'Mind, he's not your ordinary criminal. Drummond has a degree in business from Glasgow University. He has an MBA. Smart. Slippery. Violent. The only other oddity to him is ten years ago he got religion, or so they say.'

Plug raised an eyebrow. Francie grinned. Even with the front teeth in place, it wasn't a pretty sight. 'He met some young fella' when he was at university. They got real tight. He staked him in some computing business. He was even best man at his wedding. When the bloke set up some whack-a-mole church on a Scottish island, didn't Drummond go with him, along with some of his crew. Not that he's on an honest path, you understand. His people still run his operations in Glasgow, and apparently it makes for an easy life for everyone. The police don't have him in the city. His people don't have him breathing down their neck. And he gets to go with his God, although I don't know what God would have him. Leave him be, Plug. Not a man you want to meet down a dark alley or anyplace else.'

Plug nodded. It sounded like good advice. The only problem was it was already too late for North to heed it. North wasn't just heading for an island of crazy people with a bloodthirsty leader. When he got to the island, there'd also be a violent gangland boss living there, complete with his cronies. A boss who would doubtless go to any length to protect the money the Narrow Yett was helping him make.

16

North Atlantic

North's fingers found the captain's waterproof again, sliding along the wet vulcanized fabric till they found his collar.

Amid the white whiskers, the man's lips were blue and his eyes closed, his body spreadeagled on its back in the choppy water. North yelled Mac's name over and over till the eyes fluttered open. As he came to, he retched up a stream of salt water.

'Where's my bloody boat?' he said, twisting and turning in the water as a look of confusion spread over his face.

'You're welcome,' North said, but he wasn't sure if the captain heard him against the raging wind. 'Can you swim, Mac?'

'I'm swimming now, aren't I, you eejit.'

North let go of his lapels and Mac's face disappeared underneath the water then re-emerged, spluttering. There was no sign of Mac's boat, only churning dark-grey water and smashing-down rain. 'That way,' he said, chopping through the air to signal to Mac the direction he was hoping the boat went in.

'Are you sure, lad?' the captain said, a note of desperation straining his voice. 'We'll not last long in water this cold.'

His legs cycling, his arms moving from side to side, pushing the water backwards and forwards, North surveyed the sea, which looked identical whichever direction you faced.

'Fang?' he said, and pushed his index finger into his ear, but there was only silence. Mac looked bewildered, and he was right to be. If North got this wrong, they would swim until they drowned and that wouldn't be long. Still treading water, he scrunched up his eyes against the sea spray and the rain and made out something – but he couldn't be sure what. 'No,' he said, and started swimming.

The seat cushion was the first thing they found, and then another. The lifebuoy was their best find as it gave the captain the chance to rest a few minutes. It looked as if the girls were throwing out anything they thought would float. Or had the boat indeed overturned and this was the wreckage? North tried to hope for the best.

For an old man, Mac was doing well, North thought. But he was exhausted. Even though the sea had dropped and the wind levelled out, the waters were perishing and the future looked bleak for both of them. And Mac wasn't the only one who needed to rest. North was losing sensation in his own hands and his arms and legs were screaming with the effort, but every minute they rested, the boat moved further away.

He started laughing and wondered if he was becoming hysterical. Was hysteria a sign of hypothermia? Fang was going to be so mad with him if he drowned. He hoped Hone didn't drag her into more intelligence work. She should walk away and be with kids her own age and IQ. Not that he'd ever come across anyone else with her brainpower and skill set. '*I'm a prodigy*,' she once said. '*You're a moron-person*,' she said often. '*We're a great team.*' It was like she was right there with him. He just had to rest a while, he thought, as he lay on his back. His eyes on the dark, rolling clouds. Honor, he thought. He was once in the sea with Honor after the car they were in was swept away. He remembered how she'd cursed him, the gleam of her pale feet in the moonlight as they crawled back on to dry land. How she'd refused to drown out of sheer bloody-mindedness. He wished she was here now. But she'd only call him stupid, he thought. All the women in his life had been smarter than him. Why was that? Perhaps because he got himself into situations like this one? He wished he believed in an afterlife that he could share with Honor. Maybe Mac did. He hoped so.

Mac was folded over the lifebuoy, his weather-beaten face just clear of the surface, droplets of salt water caught in his beard. He hoped someone would make sure Mac's mother was buried where she wanted to be. He himself didn't care where he was buried; the deep blue sea was as good a place as anywhere and it beat the desert sands he'd once fought in. His thoughts turned to Mia – it would have been good to see her smile again, hear the rise and fall of her laugh, and, remembering it, the cold sea closed over his face like a veil.

17

London

Plug had called her to say his contact – someone called 'No Teeth Francie' – claimed that the sex trafficker Drummond was indeed a notorious gangster. That he had a legitimate interest in one of Glasgow's biggest building companies, as well as an illegitimate citywide loan-sharking enterprise and a healthy share of the drugs market. He owned three nightclubs in the city and although there had been rumours for years about his criminality, there'd been no convictions. More intriguingly, Francie said the word on the street was that Drummond had found religion. Fang didn't believe it.

Fang considered the reasons a violent Glaswegian criminal might choose to move to an island and help run a cult. Perhaps Drummond was just draining it of its ready cash through his charities scam? But it had to be more than that. Plug's theory was money laundering, and although she didn't want to encourage him, she was beginning to think he might be right. The Narrow Yett could claim it had any amount of donations come in. They could spend it on construction and supplies and Drummond could wash his money clean, especially if he was the one to provide the builders and the goods and services the islanders needed. Of course, if he was greedy, he might also siphon off the cult's own money.

Especially if Enoch left the operations side of things to his old friend. And in that case, Enoch might know all of it or none of it.

In any event, Drummond's presence on Murdo wasn't good news. For a small island in the middle of nowhere, Murdo had a lot going on. It was home to a cult that had castrated at least one if not more of its members. And now it turned out that a violent criminal was running his operations out of it. No wonder North was on his way there. Murdo was just his kind of place.

Fang had moved on to a Starbucks in Paddington station. She perched on a high stool in the window and yawned so wide her jaw cracked. There'd been a few hours' sleep last night, but not enough. Really, she should grab some more rest while North was on the boat. But she couldn't. Whatever was going on, she didn't like it. She thought about Catina's face. Her shame. How alone she'd looked climbing into the police car.

Hone, of course, had been no help when it came to Catina. Had said he couldn't interfere with an immigration issue. Which was a downright lie. Interfering was what he lived for. He just didn't want to interfere on behalf of a sixteen-year-old Romanian waif and stray who was no use to him.

As for herself, she knew the one-eyed man considered her useful, but she had no doubt that he hated the fact he had a teenager on his team. And one who was way smarter than he was. What did he expect? His generation was so out of step with the challenges of the modern world. They knew nothing. They couldn't even code. Unlike Catina, who probably could code, or even if she couldn't, would be able to learn how to.

Catina was a citizen of the world – why shouldn't she live here if she wanted to? She was brave and she had the gumption to travel to another country looking for work. She'd be a useful member of society.

Fang sucked at the tip of her plait. North felt too far away. He'd had plenty of chance to mess up without her guidance. Admittedly, she and Plug had rescued Catina from sex traffickers, but who knew what other excitement she was missing! Plus, the thought of what North was doing wrong was driving her insane – she could only guess at the mess he would make without her sage advice. She was regularly the only thing between North and disaster. Look at that debacle in the bar – it was only thanks to her navigational skills that he'd got himself onto the boat.

She checked the map again. The little blue dot currently adrift in the Atlantic, crossing over from Harris to a remote island full of conspiracy fruitcakes who encouraged self-mutilation and religious zealotry. Shouldn't he be there by now? Yep, she could think of any number of things that could go wrong there. Not least the fact North should be resting up in a hospital bed like the consultant told him – more importantly, like Fang told him. She tried to relax. He would check in as soon as he could. Meanwhile, she had no time to waste.

What else could she discover about the Narrow Yett finances?

She opened up her fake email account. Luis Moreno of the Colombian charity had replied. He was flattered to be asked to do an interview, but he said his clinic was some way off. However, he was pleased to tell her their fundraising had so far earned the best part of £140,000. And he was hopeful of more. They would do great work, he assured her. As for churches and charities, a Scottish church had generously provided funding for a women's health outreach initiative to the tune of £25,000. It had been a great success.

He was talking about the Narrow Yett. Fang sat a little further back on her stool. Enoch's cult could point to real charitable giving. But £25,000 wasn't £200,000. The figures didn't add up.

She bent back over the keyboard.

Leaving aside Tomorrow's Citizens' front for sex trafficking, over the past twelve months the Narrow Yett had channelled the best part of two million pounds into thirty-three small-to medium-sized charities – the vast majority across Africa and the developing world, many of them with an environmental, educational or health bent. The problem being firstly, donations to these charities exceeded the cult's income and secondly, only half the thirty-three charities existed. The rest, as far as Fang could make out, once she had tracked the money back, were the charitable equivalent of shell companies. Not to mention those charities that existed and were getting money were only receiving a fraction of the declared donations.

She kicked back, allowing her mind to process everything she had written and seen and learned. There were millions at stake in the Narrow Yett operation, and where there was money there were people who would act to protect it. People like Craig Drummond.

At least she and Plug had figured out enough about Drummond so North would know to avoid him.

What else did she know? She tapped out a list on her laptop.

- Jonathan Gaffney was a murderous lunatic who thought the world was a computer simulation, which was coming to an end. And the way out of the simulation would be apparent to all in...

– she checked the time –

- 25 hours and fifty minutes' time.
- Gaffney was part of a cyber cult called the Narrow Yett run by a shadowy figure called Enoch, a multi-millionaire whose wealth was originally inherited from his dead wife. She'd watched a couple of his broadcasts. 'Woo-woo' didn't begin to cover it. Although she had to admire the persuasive 'ask' to the faithful. 'What's real in this world is your generosity.'

What else?

- Enoch and the Narrow Yett were based on a remote Scottish island called Murdo.
- The Narrow Yett was a sophisticated cyber operation with two million people across the world willing to subscribe and follow their output and messaging. Despite or maybe because of the 'woo-woo'.
- The Narrow Yett was funnelling money into charities and siphoning it straight back out. That they were funnelling more money every year than they were bringing in.
- That Drummond was on Murdo and was violent, dangerous and had an MBA. That it wouldn't be beyond Drummond to come up with a money-laundering scheme to clean his own dirty money.

She pushed her Joe 90 glasses up her button nose. What else?

- That North was an idiot eager to throw himself into harm's way at any opportunity.

She snorted after she wrote that. Simply couldn't help herself.

- That Mia Barton was important to him.
- That Hone understood she would be and leveraged that fact to manipulate North into going up to Murdo when he could easily have sent another agent.
- That Hone didn't care if North died, providing he did what he was told.
- That Hone played dirty. Always.

She slipped her hand into her rucksack, found a piece of gum. The tang of artificial raspberry. The woman behind the counter glared at her. Fang adopted her most ferocious scowl and, flustered, the barista looked away.

Where to go next?

She wouldn't pretend that she wasn't still mad at North, because she was. Even that stunt he'd pulled in the bar fighting that bunch of contractors was idiotic. And he hadn't even heard her out on the money, which she'd been totally right about, by the way. But moron-person or not, she wanted him across everything he needed to be across while he was up there. Information might be the difference between him staying alive or getting very dead. Eventually, he would understand that, and understand that if Fang said one thing and Hone said another, he should, of course, listen to her – who had his best interests at heart. Rather than Hone, who didn't.

Fang attacked the keyboard again, pulling up maps and histories of Murdo.

She absorbed the geography of it, the flora and the fauna, the meteorology, the geology and the archaeology. Some called Murdo the curs'd isle, lost in the sea and abandoned by those who once lived there. It had even been left off maps till the 1930s. She flicked through archives of old photographs of hardy fishing folk and farmers and digested any number of dissertations and academic reports. Although the island was originally permanently occupied and had been for thousands of years, Murdo's residents had been evacuated in 1929, the year before the same thing happened on St Kilda. When Enoch inherited it, the island had little other than the rocket tracking station, which originally went up during the Second World War, and an old manse traditionally used for summers by his dead wife's family.

When the UK's Ministry of Defence gave up the tracking station in the nineties, D'Urberville Technologies took on the lease. Fang sat up straighter. North had wanted her to check out the contractors in the bar. This had to be the company who employed them. But she needed to double-check. She accessed the Salmon Inn's CCTV camera, input the time she was looking for, and saw four bull-necked men and a rat-like guy in aviators enter the pub. She rolled it back. Slowed it, watched again. Rolled it further back, zoomed in, slowed it and pulled up a black windcheater with the letters DUT on it within a shield.

She switched her attention to the company website. To the *New York Times* and the *Washington Post*. D'Urberville wasn't a big player in the satellite industry but respectable enough. What was more interesting was the fact D'Urberville was a subsidiary of a global defence company called Benson Corporation.

The hubbub of conversation, the tapping of keyboards from

other customers, the hiss and spit of the coffee machine faded to nothing. Fang was aware only of her screen.

Benson Corp was run by Mrs Elaine Benson, 'age unverifiable' according to the *New York Times*' fact checkers, with fingers in any number of pies – aircraft, weapons systems, space, radar and artificial intelligence. Fang was working through a virtual private network because she always did. She was tired though. Too tired, really, to do what she was about to do. But then she was Fangfang Yu and she was extraordinary.

She cracked her knuckles and caught the disapproving sigh of the barista.

'You with the laptop!' the woman called over. The voice was loud as if she wanted to embarrass her. 'You can't just sit there hogging our Wi-Fi. You should order something.'

'You should pay more taxes,' Fang said, 'and I'd be "Mocha Frappuccinos for everyone!".' She went back to her screen.

'Just do it', North had said down the line. Like it was simple. But nothing was ever simple. 'Just do it', he'd said, like he was her boss and she was the office junior. He had no idea what 'just doing it' meant. The art. The electric creativity of it. The raw physical sensation your brain was expanding and rebuilding and expanding some more. 'Just do it', he'd said, like all she had to do was close her eyes and jump into the pool. What was he really saying? That she was guilty of overthinking. That she was risk-averse. That she was a child. Fang closed her eyes. Opened them. And jumped. In the zone. Code flew from her fingertips as she worked her way through first D'Urberville and then Benson's internal systems. Firewalls turning to ashes under her onslaught. Working her way through the architecture of the site. Stripping back the layers. One after the other. Resting. Reading. Disregarding.

Searching. Going deeper. Like drowning. Till she found something.

Plans for three rocket launch pads on Murdo, interconnected with small roads. It wasn't much. It was public record that there were rocket launch pads on Murdo. And she almost didn't see it at first.

In the corner of the architectural blueprints marked Phase One, there were three scribbled letters, as if someone had signed off the plans. TIO. She pulled up the list of employees and company directors. No senior employee had the initials TIO. Something about the initials was familiar though.

'So, are you going to order something?' The woman was right alongside now, her hands full of dirty crockery she'd already cleared from the other tables.

'It's still a "No" from me. But I'm guessing you're going for "Employee of the Month". Am I right?'

The woman gave up.

A hacker called Sam99 on a board she dipped in and out of was owning key American institutions right now. Over the past three months, he'd identified ways into the systems running the Pentagon, the CIA, and the chairman of the Senate Armed Services Committee. His backdoor into the Defense Advanced Research Projects Agency had been up and down within twenty-four hours, presumably because DARPA spotted the hack. But among his offers had been a way into the Technology and Innovation Office with the US Space Force. People did that sometimes, if they found some smart way round a cyber security system. Just to prove how clever they were.

Sometimes there was a patch already in place by the time you took a look, but occasionally the alarms hadn't gone off and it was a free-for-all. It was considered a no-harm intrusion. Take a walk through. Don't take anything. Don't wipe anything.

Don't be a smart-arse. Return the favour when you've figured out a way into something other people were looking at. Plenty considered what they did a public good. They'd let their friends wander before alerting the system engineer to the holes in their system before a hostile actor could stage a break-in that was going to cost money or intellectual property or state secrets.

'The manager says if you don't order something, you have to leave or he's calling the police.' Fang had to give it to this barista. She loved her job.

'Madam, I *am* the police,' Fang said. Well, not entirely true, but close enough.

With an exasperated tut, the woman turned on her heel and stalked back to the counter.

Fang hesitated, her fingertips resting on the keys. Sam99 wasn't someone she'd had dealings with. Usually, she only worked with people she had history with. But she'd seen his name around the place for a while now, and the guy seemed to know his stuff. The thought of North walking into the unknown came to her. His admiration as she – yet again – came up with the goods. And she started hitting the keys. For fifteen minutes or so, she skulked and slid from file to file, growing more uncomfortable by the second. It was all too tame – too bland. Something about what she was seeing was off. She glanced up and around. Was she being watched? But around her, everyone's heads were bent over their tables. Even the barista had forgotten her.

Fang disconnected, covering her tracks as well and as deep as she could, bouncing from server to server and country to country. But nothing came after her.

She pinged a message through to a hacker called Lestat she gamed with.

Any dealings with Sam99? I may have messed up.

Hey, Miho. Don't know him. What did u do?

Smash no grab. TIO USSF. Felt wrong.

… Full wipe down?

Like u wouldn't believe.

Will check him out. Laterz. Hope u're good.

She hoped so too.

18

North Atlantic

It was the noise that brought a coughing, spluttering North back to the surface. Waves slapped against his skull and his ears ran with water, but even so he could make out the muffled engine and rotating blades of an approaching helicopter. With a wet hand, he tried to clear his eyes of the salt water. North had travelled in enough helicopters during his time in the army. And, if he wasn't mistaken, this one was an AW139, flying at around 1,000 feet, and it looked to be dropping. North wondered if the pilot had spotted something the girls had thrown from the boat. Hope surged through him.

It was coming in fast and low over the sea, but was it low enough to spot North and the captain? Grinning because he couldn't help himself, treading water furiously, he waved his arms, shouting at the captain to do the same. The old man was laid out on his back, one seat cushion under each arm, but at North's shouts, the fisherman righted himself in the water, and with what energy he had left, he lifted aloft one of the cushions and waved it back and forth. North heard himself shouting up towards the chopper, though the logical part of his brain knew there would be too much noise in the cabin to make out their shouts for help.

The twin-engine helicopter rose higher in the sky again, banking, and then coming round. North shouted louder. They'd spotted them. Thank God. The 139 hovered over them, coming down low for a better look till it was no more than ten metres directly above them, the force of the downwash pressing against the sea, rippling it further and further outwards and throwing up a fierce and stinging sea spray. The copter moved to one side and a door rolled open.

Scrunching his eyes up and using his arm to shield them from the downdraught, North squinted upwards towards the open door. It might as well be a mile away. There was no way to reach the helicopter and it wouldn't be able to come in much lower, would it? The pilot couldn't put the lives of his passengers at risk for them. Was there any chance at all that they had a rope in there and that the passengers would be able to haul them up, each in turn?

A dark figure in dark sunglasses and ear protectors leaned out of the open door. Then another, and another. Something familiar about the way they sat. Did they have a rope? Could he be that lucky? He watched the sky for the snaking rope to appear and to fall. Were they arguing among themselves as to how to winch them up? He would slip the rope over the captain's head and arms so that it caught under the arms. It wouldn't be ideal. But it would be good enough.

No rope.

Then he remembered. This was a top-of-the-range commercial ride, which meant that if travelling over water, the copter would be required to travel with an inflatable life raft in the event of ditching in the sea. Was the delay down to the fact they were accessing the life raft? A raft would still leave them out at sea, but the pilot could ring in their coordinates to the coastguard and the storm had dropped away, so they should be good until help came.

He glanced across at the captain to warn him to be ready. That the life raft would inflate immediately on impact with the sea. He didn't want the old man knocked senseless by it. He wanted that life raft more than he had wanted anything for a very long time. The idea of being out of the sea and able to rest felt like the closest he was ever going to get to heaven. Mac seemed ready and the helicopter was side on to them now and a little bit away so they should be safe enough from the weight of the falling life raft.

North looked back up. The man in the open doorway was grinning. A toothpick between his startling white teeth, ancient red-and-white burns across his face. Kincaid. The guy from the bar. And North knew that there was going to be no rope and no life raft.

Instead, Kincaid flipped the two drowning men below him the bird as the helicopter rose, banked again and pulled away from them. A fading sound of whooping and cheering. The helicopter lifted higher in the sky and North watched as it disappeared into the distance. He let out a roar – equal parts frustration and rage. Three hundred miles per hour, the thought came to him. Couldn't get away fast enough now they'd decided to do nothing.

The helicopter was transporting the contractors back out to the island. They hadn't used the airport, but had to have taken off somewhere close to Auchterburgh. He'd humiliated them, beaten them and escaped them. He imagined the contractors together in the cabin laughing, the noise of that laughter through their headphones, vibrating in the bones of their skull till it was all they could hear. Imagined the kind of ugly comradeship that they shared. To make a decision to leave two men to drown. He should have hit them harder.

Alongside him, the captain swore at length and with impressive creativity. Their abandonment apparently reviving the old man.

Perhaps they would think better of it and the helicopter pilot would call in their position and the coastguard would come and pick them up. That would, after all, seem the least they could do. But North knew that wasn't happening.

'Did you jump in after me?' The captain's voice again, furious with North now. 'Did anyone ever tell you that you're one stupid bastard?' He guessed Mac didn't want North's death on his conscience. Or perhaps he was remembering the threats of North's pursuers on the quayside. Had he made the connection? That whatever North had done to the men back on Harris, this was pay-back time.

'You have no idea,' North said.

'We're done for, lad. You should have left me to it and stayed safe.'

North shook his head. If his first emotion had been anger, his second had been an acceptance that he wasn't in this alone and that the old man would need him more now because they were the only ones who could get themselves out of this mess. They were comrades under fire. The passengers' callous disregard had revived North. He had no intention of drowning in the middle of this sea or any other. He was getting Mac out of this mess and then he was finding Kincaid and breaking each and every one of his fingers before he killed him. Not because he was angry, but because that promise to himself was going to help him survive this.

The wind had dropped – at least there was that. The

helicopter had waited for the squall to pass before it transported the contractors out to the island.

'How far are we from Murdo, Mac?' North said.

'An hour and a half by boat. Swimming, we're never going to make it.' North set his jaw. Mac was wrong. They were going to make it or die trying.

19

New York

The old woman's hands showed her age. Not even the glittering diamonds on the twisted fingers could cover the roped veins, but they did at least distract her from the pain. She flattened her hands as much as she was able to on the desk in front of her and watched the rainbow lights from the precious stones dance across the print of the *Wall Street Journal*.

'Mother?' The questioner spoke in a deferential tone.

She raised her eyes to his, brought her hand beneath the desk and flexed her stiff fingers. It was five thirty in the morning, but even so her make-up was flawless. 'You told me it had been dealt with, Teddy. That the problem had "gone away" with the death of that little man. You said we were back on track. I don't want to have to tell the generals differently.' The gnarled fingers went to the star-spangled banner brooch of sapphires, rubies and diamonds set in white platinum that she sported on her cream cashmere robe. 'I do so hate making them cross.' Her eyes were an arctic blue, in contrast to the charming smile on her carmine lips. 'I can't pretend I'm not disappointed.'

Mrs Benson had fostered Teddy when he was three and legally adopted him the year after. 'I chose you out of all the boys and girls in the world,' she had told him throughout his childhood. 'Because you are so very special.' He was also biracial, startlingly handsome, with the highest IQ the care home had ever come across. She put him through the best schools and gave him all that money could buy. But Teddy had always known there was a price to pay – unquestioning devotion. Fortunately, he was more than willing to pay it.

And she did love Teddy. Not as much as she loved her country and not as much as he loved her. But he was her legacy and she certainly loved him a great deal more than she had loved her husband, Henry Adolphus Benson III. Henry was long since dead. An unfortunate dive into a shallow lake in the grounds of their Italian villa, all while horribly drunk and not yet thirty. But nearly forty years later, Mrs Benson – the majority shareholder in the Benson Corporation – was still very much alive.

The woman was a legend in the markets, steering the company through four decades of war and technological revolution. Rumour had it she had killed her mother on emerging into the world. Who knew the truth of it now? Rumour also had it that when her young husband had his unfortunate accident, she had slipped into the bed of his grieving father and never left it. A happening so offensive to her mother-in-law that divorce followed within three weeks. Everybody in her past was dead now and no one left to tell the tales. The only fact of the matter was that Mrs Benson was one of the richest women in the world. That she was a friend to presidents and celebrities alike. That she was one of New York's most generous philanthropists and had been responsible for the deaths of thousands of soldiers and civilians across the world, courtesy of the company's cutting-edge weaponry. That in her hearing she was called a true patriot, and out of her hearing 'the Bloody Widow'.

She was still a beautiful woman. That much was clear in her even features and fine skin. And she could afford the best surgeons. Her hair, too, was still a teased-out honey blonde, kept thick and glossy with extensions that had once belonged to women less fortunate than she was. The years had taken their toll, of course, but money lent her a lustre. The Chanel suits helped, the Kelly bags, the oversized pearls. Most of it, though, was for the benefit of others. The uniform of the uber rich. Armour to go out in the world among them. But it had never been about the money for her. Money was merely a tool. Something that helped you win. And Mrs Benson needed to win in any game she played, because power was her oxygen. She would die all too soon, she knew that. Ten years, maybe twenty if she was lucky. And when she died, the world could end as far as she was concerned. It could burn – everything in it except America. Everyone in it except Americans and then they could start over. In fact, she would prefer it if they did – under Teddy's leadership, of course.

She brought her hands up and settled her elbows on the desk, leaning towards her son. 'For generations this company has helped to keep this great country safe. No one can be allowed to get in the way of that, my darling. If only I wasn't so frail and so feeble.' She had the constitution of an ox and they both knew it, but even so she sat back in the leather chair with a sigh.

'Alan, my coffee and grapefruit.' She waved her ringed hand towards the far end of the room and out of the shadows slid the man who had served as her butler for the last thirty years. He wasn't called Alan. She didn't care. She stood up from the desk and tottered across to the cream brocade sofa where two white Shih Tzu dogs waited for her, settling herself down next to them. One dog was called Ronnie and the other Reagan, after Mrs Benson's favourite president. She'd had a Trump, but she had it put down after it nipped her.

Mother didn't say goodbye but Teddy knew himself to be dismissed. He would do what was necessary – he refused to let her down. He hadn't even told her the worst of it, he thought, as he let himself out of the apartment. A lone wolf on Tower Bridge was one thing. They could batten down the hatches for a while. A grieving sister with awkward questions was unfortunate, but could be handled.

What he hadn't told his mother was that their contacts within the US Space Force had also been on to him. There'd been a hack and the hacker was trawling for info on Benson's interests on the spaceport at Murdo. Now that was bad. Something would have to be done, which meant calling in favours, but that was okay because Teddy Benson had plenty of friends where it counted. But even then, Teddy had worse secrets. At least the worst thing was buried deep. And the hacker was looking in the wrong direction when it came to that at least. But for how long? Teddy clenched his fists and the knuckles whitened. He did so hate to disappoint Mother.

20

North Atlantic

They swam and rested in equal parts, trailing the life buoy after themselves when they swam. The sea had died down but it was still rough enough that North thought there were times when the current took them backwards rather than forwards.

Mac was exhausted. 'You're going to have to leave me here, lad,' he said at one point. 'It's not the worst way to go.'

North spat out a mouthful of salt water as he eased the lifebuoy over Mac's head and shoulders, pulling his arms out over the ring – the old man's hands blue with cold. 'You and me are in this together, Mac. You lost your cap when you went in the sea. Save your breath and think about the new one you'll buy when we get out of this.'

Mac managed a chuckle. 'Right you are. I'll be getting the best damn cap money can buy.'

North hauled the ring after himself as he moved through the water. With each stroke feeling Mac get heavier, and himself get weaker. The cold had him now, he knew. And it wasn't letting go. No sensation in his legs. Trusting that they were moving back and forth. One more stroke, he told himself. One more minute alive in this world. He didn't know how many minutes he had left, but it wasn't many when he caught sight of the fizz and a pop of first one flare and then another.

Was it his imagination?

Was he drowning, half-dead, and dreaming?

A grey spot appeared on the horizon. It grew larger. It was the boat and it appeared to be drifting. He shook Mac back into the land of the living.

North yelled the girls' names and the captain did the same. Were they on board? Had they too been swept out of the boat by the storm? From some deep reserve he didn't know he had, he found the energy to start swimming again, breaking off every now and then to yell some more.

The sight of Comfort leaning over the side, her blue hair whipping from side to side as she waved and hollered out to them, was the best thing he had seen for a very long time.

She said the engine had cut out when a huge wave slammed into the boat shortly after North jumped into the sea. They had been in the sea an hour – it had just felt like a week. North knew Mac wouldn't have lasted much longer. The old man took time to warm up. His teeth still chattering in his head, he disappeared down below amid curses and bangs and shrieks of metal. Eventually, he climbed back up the wooden stairs, a look of steely determination on his face. Rubbing his calloused hands together in a bid to restore circulation, he met North's eye. If this didn't work, they weren't going to make it.

He pulled out a small bottle of whisky and held it out to Joy.

'You shouldn't drink alcohol when you're cold,' she said. 'It's dangerous.' Then she gave him a broad smile. 'Then again, it seems rude not to.' She unscrewed the top, took a mouthful, then held out the bottle to Comfort, who shook her head. North took it from her. 'To the confusion of our enemies,' he said, raising the bottle. He meant Kincaid.

Mac nodded, and took the bottle from him. He took a long drink, wiped his mouth, sighed then took hold of the wooden wheel. Shaking his head as if in disbelief that he was still alive and back where he belonged, he blessed himself then turned the key in the engine.

North held his breath. He trusted in Mac. The girls clasped hands and trusted in whatever God they believed in.

Nothing.

Again.

Nothing.

Mac closed his eyes as if in prayer, turned the key and the engine shuddered into life.

The engine had flooded as the storm blew itself out, according to the girls. In quieter seas, they'd initially attempted to steer the boat back round to where they thought North and the captain had gone into the water, and Comfort said it had been Joy's idea to throw everything they could think of into the sea both as buoyancy aids and as a way for them to follow their trail, including seat cushions, the captain's flask, two lifejackets, and the lifebuoy. She nudged Joy, beaming from ear to ear, when North said they'd found the cushions and that the lifebuoy might have saved the captain's life. Increasingly desperate, the girls had tried raising help on the radio but to no avail and instead they dug out four pyrotechnic flares. One had been a fail, they'd used up two and there was one left.

'Shall we send the fourth one up?' Joy asked, but the captain shook his head. 'Only if we see that helicopter come back over, lassie, when I'll be firing it at them.' North didn't think the old man was joking.

'The men you wouldn't take the drinks from bear a grudge,' North explained. Comfort nodded.

'We tried waving at them but they didn't see us,' Joy said.

No chance, thought North. They had spotted North and the captain in the water. They would have been watching for a boat or the wreckage of the boat. North had been angry before but at the thought the men had been so callous as to leave the two girls struggling on an unseaworthy boat, he was ready to tear someone limb from limb. He allowed the fury to flow through him, to take hold of every part of him, and then he let it go. Out here in the middle of the sea, he had more pressing concerns. Like making it back to dry land.

Joy handed him a dry towel she'd fished out from some drawer. It was threadbare and moth-eaten. It might dry his face if he was lucky. North guessed Mac didn't keep many creature comforts on the boat. 'I'm sorry I was rude earlier,' she said in a quiet voice. Two spots of pink bloomed on her cheeks. 'You were fated to be on this boat so you could save Mac. The Universe put you here. I get that now.' She gave him a serious kind of smile, and he nodded back as if he'd thought the same thing. And now they all had to hope that the Universe didn't want them to die at sea.

Should he try and make a call? His gaze sought out his rucksack – it was unbuckled. Comfort followed his gaze.

'We searched it. I'm sorry,' Comfort said with a blush. 'We thought you might have a mobile phone we could use to call for help. I'm guessing you had it on you and it's at the bottom of the sea now?'

He made a play of patting his pockets down, then nodded as if grieving the lost phone.

In reality, the satphone was secreted in a special waterproof pocket at the base of the rucksack – the P226 along with it. He opened his mouth. He was a busted flush if he had to get out the Iridium and use it to call Fang. It would mean the operation was over before it had begun. He closed his mouth again. Watching the captain, he didn't think he needed to. Mac was drenched and exhausted, but he appeared to be feeling his way to the island by an instinct honed over years at sea.

'You'd tell me if we're heading for the edge of the world, right?' he said, pulling on dry clothes from the rucksack. He felt Comfort's eyes on him. Mac nodded. It had been a joke, but with only the grey sky and the grey sea around them, North felt the truth of it. Enoch had settled a long way from civilization. Legacy or no, he wondered why that was, while overhead seabirds wheeled and banked, sinking and rising on the breeze, one or two dropping down to the sea, skimming the surface for fish. He watched as one rose again, a still-struggling fish clasped tight in its talons, drops of water cascading from the silver scales back into the churning sea.

'Land ho!' Joy let out a shout as she caught a glimpse of a Zodiac coming up on the starboard, and then another. Two more Zodiacs and a motorboat were skimming the sea, heading straight for them. Shouts and halloos went up.

'They came looking for us.' Joy turned to Comfort, and the girls hugged each other.

The Zodiacs slowed as they reached the boat. Their various crews called up to the girls and started whooping and waving.

Both girls waved and 'halloo'd' back. They seemed beyond pleased to be home. North wondered if they knew how dark and dangerous home could be.

21

London

The number 23 bus between Paddington and Exhibition Road smelled of other people's bodies, last night's takeaways, and London grime. Fang was looking forward to seeing the neuroscientist and computer nerd Dr Paulie Holliday. He was one of the few people in the world she considered almost as brilliant as herself. His usual area of expertise was artificial intelligence, but AI was a huge topic in the world of defence, as she'd recently found out to her own cost. As a result, Paulie kept track of all the big players in defence these days. Not least because he didn't trust any of them. If anyone knew what Benson might be up to on Murdo, that person was Bald Paulie.

Fang didn't hear the woman at first – just became aware of the blossoming smell of fresh sweat. Fang glanced down the aisle at the approaching passenger. A six-foot runner in a hoodie and dark leggings that accentuated her athletic frame, thirty-something with an asymmetric blonde bob and overplucked eyebrows adrift in a frozen forehead. The bus stopped, its brakes squealing, and Fang glanced down at the waiting passengers at the stop. Marble Arch. People jostled and pushed around the doors, but no one was getting on. A hard case banged on the glass as Fang used her finger to pull away her earbud.

'Young lady, may I have this seat.' The runner's accent was nasal Bostonian, and it didn't sound so much like a request as an order. The runner pointed at the empty seat next to Fang and, as she did so, Fang realized the forearm was more slender than it might have been and that the woman's left hand was tiny compared to the right, made smaller again by the fact the ring finger was fused with the little finger.

Lifting her duffel bag, Fang glanced around the upstairs of the bus. There were plenty of other seats. In fact, every single seat other than her own was empty, which was strange because a couple of stops ago, there'd been around a dozen passengers. Weirdo alert!

And what was happening with the bus? It wasn't moving. Had it broken down? But even as she thought it, the bus jerked and juddered, as if it had pulled back into the stream of traffic, and there was a blare of horn from behind it.

Fang scowled at the woman, before slapping the bag back down on the seat. She wrinkled her nose to let the woman know she smelled bad. 'Sorry.' She made sure not to sound at all sorry. 'I'm saving it.'

'For a friend? I've read your file.' The tone was dangerous. 'You have no friends.'

Fang felt a flicker of what might have been classed as fear in someone else as the runner moved the duffel and dumped it out on to the aisle. Anyway, the woman was wrong. She did have a friend – North. In fact, she had two, because there was Plug as well. Three, including Bald Paulie. And an AI system called Syd, which also claimed to look forward to her visits. It currently lived in a convent of cloistered nuns. Which made four. So there. The woman sat down, her damp left thigh and shoulder pressing against Fang's. She shifted nearer again, pressing the girl closer to the window than she wanted to be.

Fang's gaze locked on to the malformed hand – like that of a broken doll, the fingernails painted with tiny American flags. 'You Brits are meant to be polite. Didn't you get the memo.'

Fang looked up into the woman's face – made sure to keep her face neutral as if to match the woman's. Whoever this stranger was, she was trying to intimidate her, and Fang wasn't going to give her the satisfaction of knowing it was working. She kept quiet. Naturally, she wanted to know who the woman was and what she wanted, but she refused to ask. Instead, her own hand slipped into the pocket of her black biker jacket and she felt the woman next to her stiffen. Fang pulled out her packet of gum and popped in a piece. She started chewing, making it sound loud and juicy, before turning to stare out of the window at the passing buildings.

The runner sniffed her disapproval at the gum or at Fang – the girl wasn't sure which. 'You messed up, Fang. I can call you Fang, right? And you can call me Roxie. You think you're special, but you're not. You think rules don't apply, but I'm here to explain that they do.'

Fang was pretending not to listen. But she was listening hard. 'And when you break them, guess what – there's a price to pay.'

Under her breath, Fang swore. She'd tripped a wire, she knew it. The only thing that really surprised her was how quickly they'd managed to track her down. 'Just do it.' That was some genuinely terrible advice from North. He was a complete moron. What had she been thinking?

The American was watching her work it out. She smiled at the girl, her white teeth smaller than they should be. Like milk teeth, Fang thought. As if part of the woman's DNA had never wanted her to grow up.

Fang felt her own anger rising. She couldn't believe her

stupidity. She hadn't just tripped a wire. They'd been waiting. It had been a trap and Sam99 a front. He'd had too much of the good stuff. He'd floated around for a year getting his name around the place and then it had started. A backdoor here. A suggestion there. Building up trust among those most reluctant to trust. The Pentagon's own cyber warriors had been trawling for hackers. Inviting them in with a welcome mat, just to snare them in a net and then scoop them up. Fang thought about Granny Po, sitting in the Bowes Lyon Suite at the hotel watching daytime television, knitting and waiting for her to come home. She thought about North in the middle of the sea, heading for trouble. Would she be able to get a message to Hone? And she had to get a message out to Lestat to shut Sam99 down before he suckered anyone else. This woman was from the Pentagon or the CIA or the NSA, with all the power and the might of the state behind her. She also smelled like an old gym sock.

'What's going to happen next,' Roxie said, 'is we put you on a plane to the US. We'll fly out of a US airbase and there'll be no witnesses. When you get off that plane, which will land in a US airbase, you'll be taken to a private court and sentenced to lifetime imprisonment as an enemy of the United States. Your age will count for nothing, other than to make you particularly vulnerable to a certain type of predator in whatever dark hole they throw you into. When your government asks where you are, we'll already have you in our system. They won't ask twice, because ultimately, Fangfang Yu is a nobody.'

Fang started to blow a blue bubble, smelling of synthetic raspberries, and with her good hand the woman punched her hard, once, on her right cheekbone and her left temple slammed into the window, dislodging her glasses so that they almost fell from her face. Her head rang with the shock and tears sprang to her eyes, but she refused to let them fall. The woman carried on as if nothing had happened, other than the small doll-like hand patting Fang's knee as if to reassure her that it would never do such a thing. As Fang settled her glasses back on her nose, she thought about North, but he would never get back in time. Plug would be on her side, but he was an amateur in this game. Though Edmund Hone would help her, surely?

'Hone will be angry, you're right.' It was as if Roxie could see into her brain. 'And he does have influence. But he also has superiors and they've already been squared, so there'll be nothing he can do. Sure, he'll make a fuss among those friends he has in the US with influence, but eventually they'll be instructed to stop taking his calls. Hone's a pragmatist. Sooner than you think, he'll reconcile himself to the fact you're gone, and he'll forget you.'

To her shame, Fang realized one tear had made its escape. It cooled on her swelling cheek. North, she thought, would not forget her. He would come for her, but he would have to find her first and there was always the chance he would be killed for his trouble. Plug too. Before that happened, she would kill herself, she decided. At the first opportunity. She wouldn't let them win and she wouldn't let North and Plug get themselves hurt on her account.

Roxie was waiting for her to run through her options, she realized. Which were nil. She was about to go to prison – a prison a long way away. And one she was never going to escape.

'There's an easier path,' Roxie said, standing. The doll's hand went to the bell on the pole and pressed it, twice. 'You stay away from D'Urberville Technologies and Benson. A long way away. They're good people. And you agree to work with us. That way, when we ship you back to the US, it's to a job commensurate with your abilities. You belong to us – get used to it. Right now, you're thinking "never in a million years". That's how it always goes. I've seen it over and over. You're a clever girl, I'm told, or I wouldn't be here. We want eyes on North. Don't think of it as betrayal, think of it as a way out. You're young and adaptable. And even if you weren't, trust me – treason is surprisingly easy. I'll be in touch.'

A way out. Fang despised herself for the lightning lift of her spirit.

22

Roxie struck Fang as a deeply unpleasant person. And not someone who made idle threats. She'd made it crystal clear Fang was to abandon any attempts to gather more information on what Benson was up to on Murdo. The sensible thing then would be to go nowhere near the topic of Benson again. Then again, Fang was a genius, so a sensible course of action was never going to be her thing. Plus, there was nothing Fang liked better than doing something she'd been told not to. Especially when whatever it was that Benson was up to had drawn the full force of the US government down on Fang's head.

She got off at the next stop. Coincidentally, it was close to the Empire and she was tempted to call Bald Paulie and cancel then crawl home. But she wouldn't give Roxie that satisfaction.

Though she didn't want them following her to her rendezvous either. She watched shop windows for any sign of a tail. They were good but they thought they were following a child, so she figured they were less careful than they might otherwise be. In fact, she was counting on it. She dipped into a French patisserie and took care to stand by the window in plain sight as she bought a box of chocolate macarons from the girl behind the counter. As she waited for her change, she messaged Lestat:

Sam99=state goon

Along with a skull and crossbones emoji. Lestat messaged back with a sad-face emoji and then:

Understood. Good luck, Miho.

It wasn't much. But they wouldn't be able to play the same trick again. Lestat would make sure of it. When threatened with years in a prison, funny what you think you'd miss. Sure, Granny Po, her mother, North and even Plug. Fresh air and sunshine. But raspberry-flavoured bubblegum and video games featuring trolls and orcs and giants would be up there.

Stepping outside the patisserie, she hailed a cab and instructed the driver to drop her off on Brompton Road. The crowds of shoppers and tourists were just beginning to build. It took her more than an hour before she was confident that she'd given the tail the slip, walking into and through the Victoria and Albert Museum's shop, the sculpture gallery, the courtyard and out the exit to Exhibition Road, losing herself in a school party, just one more teenager among many.

As she dipped and dived, she contemplated what it would be like to work in an American cyber weapons factory. Because that's what Roxie meant when she talked about 'a job'. Fang wasn't naive. She knew the US needed digital defences to counter cyber warfare waged by countries such as Russia, China, South Korea and Iran. But she also knew that whatever weapon one hostile nation could use against the US, the US would want something even more powerful in its own cyber armoury. However Roxie dressed it up further down the line, Fang would be working on projects to shut down hospitals and power plants, to bring down civilian aircraft and create financial chaos in the markets. Casualties and mortality rates would be gamed. Until the day came when they moved off the spreadsheet and into the real world. And innocent civilians would be the ones to pay with their lives. Not innocent civilians in America, but in any and all countries the US considered a threat. Fang shuddered at the prospect. That was the real alternative to years spent in the American penitentiary system. But at least she'd be able to order hoisin duck pizza with extra anchovies and feel the rain on her face. If she could live with the guilt, that is. Because she wasn't fooling herself. Her freedom was at stake. More than that. Her moral code was at stake. If she didn't cooperate with Roxie, she was going to prison. And if she did cooperate, she was going to hell.

By the time she got to the Space Gallery at the Science Museum, she was in possession of a very bad mood, along with three new phones, a tablet and a laptop. She had to assume that the Americans were across the old ones. She gazed up at the Black Arrow R4 rocket and the small satellite hanging close to its open nose cone.

'A testament to our lack of ambition,' a voice said next to her. Bald Paulie smiled sweetly as he turned his gaze on her, his alopecia and plumpness giving him the shiny damp look of something newly born. He gave her a squishy hug and Fang felt her mood lift. Bald Paulie had that effect on her. How bad could the world be with someone as sweet as Paulie in it?

Fang had first met Dr Paulie Holliday a few months ago when he'd helped to save her life. He hadn't got any thinner and he hadn't got any less bald since then. She guessed he had worn the NASA hoodie in honour of the nature of the interview. Usually, he stuck to a Harry Potter theme.

'Thank you for waiting,' she said.

'Are you kidding? I love this place,' he said. Bald Paulie was the smartest person she knew, aside from herself, of course. He currently held a job as deputy director of the Sullivan Hawke Foundation for the Study of Advances in Artificial Intelligence, specializing in all things defence related. If his first love was machine learning, and his second Harry Potter, space came a close third. But not even space could distract him from the damage to Fang's face.

His own face, with its lashless eyes, crumpled as he tapped his cheekbone as if it had begun to hurt in sympathy with hers. 'I thought you were supposed to leave the knockabout stuff to North, on the grounds he's horribly good at it and the size of a building. What happened, Fang?'

She scowled at him and realized scowling hurt, which made her scowl some more. The last thing she needed was sympathy, and, somehow, she thought, brave though Paulie undoubtedly was, he'd be even less use than Plug against the massed ranks of an American goon squad.

'Are you in trouble?' he said, suspicion in his voice.

Yes, she wanted to say. But didn't. She fought back the sudden urge to tell Paulie about Roxie punching her. About the threats to her freedom. About how dumb she'd been.

'Did someone hit you, Fang?' Despite his equitable disposition, Paulie's fists were clenched and he was beginning to look like someone who wanted to hit someone back. She knew the fact that she'd been hurt would be a problem for Paulie. His girlfriend had met with a grisly end and Paulie still struggled with it.

'Absolutely not,' she lied.

Paulie seemed to sense that was an end to it, although he didn't seem happy about it. He continued to stare at the swelling on her face. As much to distract him, she started talking. 'North is on the island of Murdo, investigating a cult up there.'

Paulie made an interested face, as if to say nothing about North would surprise him.

'While he was trying to get onto the island, he came across a bunch of guys from a company called D'Urberville Technologies. It's in the satellite business. It turned out that they're working on the island.'

Paulie, she had to say, was an excellent listener. His round shiny face was rapt. As if Fang was the most fascinating person he'd ever had the privilege of a conversation with, and she felt herself unwind a little at least.

'I looked into DUT. From what I can see, it's part of the Benson Corporation. And Benson is big into defence.' Paulie started nodding up and down vigorously. So he'd heard of them then. She hesitated. 'I had a rootle around their system.' Bald Paulie clasped his hands and raised his eyes heavenwards like a priest hearing about a sin. 'And found the initials "TIO" on the construction plans for the rocket launch pads. That is to say, the US Space Force's Technology and Innovation Office has an interest in a spaceport up on Murdo, which is a long way from Kansas.'

She could have added: *And they're furious I know about it.* But didn't.

Paulie waited to see if she'd finished. 'That in itself isn't surprising. The US can't dominate space without commercial partners. It's way too expensive. And satellites are big business right now.'

He pointed at the rocket above their heads. 'The UK should have sent that very rocket hanging up there into space. Instead, we had one successful launch and then cancelled our Black Arrow rocket programme in '71. We left it to the Americans to take our satellites up into space.

'It was a mistake. According to the United Nations, there are more than 7,300 satellites up there already, with plans for thousands more. More than 3,300 are active, the rest inactive. Most of them are communications satellites, some of them for weather and earth observation, or navigation and positioning, some are strictly space observation. How we live today is dependent on satellite technology – the phones we use, the TV we watch, the maps we use, whether we carry an umbrella. It's certainly hard to imagine how you'd fight a war these days without satellites in the mix. In war, they let you see what's happening on the ground, like troop movements on a border. They provide secure comms, data, logistics. What's different at the minute, in terms of the technology at least, is that the satellites are getting smaller and you can put more of them up there when you launch.'

'Why are the Americans so invested in Murdo?'

'Why not? Anyway, it's not just Murdo. There could be half a dozen spaceports in the UK within a few years. With rockets that come down in one piece as well as go up, even with aircraft. The satellite business is booming, and everyone wants in.'

He'd started walking and Fang went with him. It was like following a walking, talking egg, she thought. Wide-eyed, a small child gazed up at Paulie from his pushchair as if thinking the same thing. That Bald Paulie should never sit on a wall and have a great fall, because it wouldn't end well.

Oblivious, Paulie held up his phone and showed her a map. 'Space Hub Sutherland,' he said. 'They're looking to build on the A'Mhoine Peninsula on the north coast with a dozen rocket launches each year. Then there's one planned in Shetland on the island of Unst and another with horizontal take-off at Prestwick airport. Cornwall has its own scheme and there's even one planning to launch a bulk carrier off the coast of Wales. The idea is they get planning permission, they get built, they get licences, and they get thousands of small satellites into space. The UK plans to dominate the small satellite industry across Europe. I'm intrigued they've let the Americans gobble up the opportunity on Murdo, but it's true that we're in a global economy and DUT has its headquarters in London.'

Fang had pinged him the plans she had swiped from Benson earlier that morning. Technically, she thought, she was making him an accessory, but she'd warned him of the need for security and knew he'd take his own precautions. Bald Paulie stopped walking to bring up the plans. She noticed he had automatically used a VPN – a virtual private network – to avoid the curious being able to track his movements online. 'What's interesting is that across the UK, D'Urberville's operation on Murdo is the first to be built and licensed. They've had a test flight but they aren't expected to kick into gear properly until later this year.' He flicked to another screen. 'This is a satellite image over the island. I've a friend at Imperial with an interest in geo-observation. Look there.' The tip of his finger pointed at a brown mass. 'That is the first new launch pad. And here.' He moved the image across the island. 'And here. They're the three launch pads that you sent me plans for. You've checked for permissions?'

She nodded.

'I'm not a lawyer and you say it's the MOD leasing the land on Murdo and then subcontracting to D'Urberville? They're managing to keep the Scottish government out of it somehow. But I can say that this set-up is in place much faster than any of the other UK operations. I'm guessing that's because the permission for the build and its operation was rushed through. The question is – why? What's special about this particular spaceport? Because whatever the Americans are doing up there, the British government is on board for it. And the other question is...' He pointed to a long building away from the rest of the station. 'What's that? Because it isn't on any of the official documents.'

23

Lucy had two carers in a day. One in the morning and one at night. Another came at some point during the week to give her a shower. She had stopped trying to remember their names. Her favourite was an overweight blonde with breath like an ashtray, whom she privately called 'Ash' but was actually called something beginning with 'J'. Possibly Julie. Another was a pompous man with a white beard, who bored her silly. She dubbed him Boring Bob after an uncle she and Laurie once had. There was a girl with a greasy ponytail who talked to her like she was a simpleton, and three harassed-looking women from various Eastern European countries, one of whom had young children. But she could not remember which one.

They were supposed to come between allotted hours. In fact, they often came ridiculously early at night or ridiculously late in the morning, blaming traffic or a poor old dear who'd fallen or staff shortages. She never minded. Time meant little to her.

Occasionally, they failed to turn up at all, but she never queried their absence. Equally, she made allowances for their occasional impatience with her. She imagined it was testing to deal with someone with as little memory as her own. She had once written all their names, along with their characteristics, down in a notebook. Unfortunately, she could not find the notebook.

Too often, the agency would send a new carer round, which she found more distressing than she should. And if she thought about it rationally, she supposed in a way, all of her carers were new to her each and every visit.

This morning, she'd woken in something of a daze, as if she had been up all night. Normally the carer brought in a cup of tea, opened the curtains and helped her get dressed. Mornings were usually her best time. As if her brain had rested and was prepared to give her a few minutes of something that almost passed for clarity. This morning, however, she felt dull and heavy, her limbs weighted down with anxiety and forgetfulness. She turned over in the bed and went back to the oblivion sleep brought.

When she woke again, instead of getting out of bed, she lay petting the three cats that had chosen to sleep with her and which were stretched out on the bed alongside her. These days they all answered to the name 'Cat', though one preferred 'Here Puss'. She could tell by the traffic noises that it was late. Even so, the effort involved in rising almost defeated her. But there was something she had to do. Although she couldn't remember what, and a sense of panic began to rise in her, until something made her reach for her notebook on the bedside table next to her.

Do NOT panic, Lucy, she read, in what she believed was her own handwriting. *You have a terrible memory these days. It's okay. Really, it is. You have carers but sometimes they come late or not at all. You have left yourself notes around the place to remind you what to do. Your glasses are on the bedside table, put them on.*

This was like a game, Lucy thought. You had to follow the rules or you would lose.

She picked up the pair of rainbow-coloured glasses by the

side of the bed and the world slid into focus. She kept reading.

Put on your watch. This is an Apple watch. It will help you find your way.

She picked up the large-faced watch and did up the strap.

In this room is a wardrobe. It is marked 'Wardrobe'.

Her hand still over the rise and fall of the ribcage of the nearest cat, Lucy looked across the room from the bed. She knew what a wardrobe was, for God's sake. First horror and then amusement swept over her. Would she wake up one day and be unable to remember the difference between a wardrobe and a chest of drawers? Would she look at the cat and wonder 'What is that?'

Get out of bed and get dressed, the notebook read. *Keep this with you at all times.*

She noticed the notebook was on a small cord and slipped it around her neck as she pulled back the duvet, careful not to disturb her furry companions. They were more awake than she was, she thought. Their tails swishing backwards and forwards, their eyes on her.

She crossed to the wardrobe, opened it and stood looking at the clothes. Jeans and— She searched for the word. Jeans and—

She reached for the notebook and opened it. There was a drawing of a pair of trousers and a jumper. A bra and knickers. Socks. Each with the word underneath.

Do NOT panic, Lucy, she read again.

Remove nightgown.

As quickly as she could, she got dressed.

One shelf of the wardrobe held four more pairs of glasses, each one with different frames.

Put on glasses, the book said. *As many pairs as you can find. This is so you will still be able to see if you take them off and put them down somewhere. Do it.*

Lucy swept another pair of glasses onto the top of her head, where they held back her silver curls. The third pair she tucked into her T-shirt. The fourth came with a beaded necklace, which she settled under her notebook. The fifth were already tucked into a brooch, which she slid into the wool of the green jumper. She checked them all. Five pairs should be enough for anyone, she thought. Did she look foolish? Probably. Did it matter? She thought about it and decided that it didn't.

As she left the bedroom, the three cats leaped from the bed and padded out with her. She wondered what they were called, and settled on Kitty. No cat could object to that.

She passed a door to the left, and read the word BATHROOM in capital letters. And ENTER.

So she did. On the mirror was another white sticker, which said *Use toilet. Wash hands. Clean teeth. Wash face. Dry face.*

She did as she was told. In the mirror, her fingers found the pendant at her neck. It was in the shape of a full moon; she could see the craters and seas picked out on it in exquisite detail. She pulled it out, unclasped it and gazed at her brother's smiling face. Laurie.

'What's happening to me, Laurie?' she asked. 'I don't like it.'

Perhaps Laurie would be downstairs? That would be nice. They could play a game of— Something. Something. Backgammon. She found the word and felt a rush of pleasure.

She heard the opening and closing of drawers in the kitchen. The rustle of papers. A familiar fear settled around her throat, the beat of her heart picking up, panic building. Was someone here? A thief? Was there a thief in her house? In the hallway, her fingers found the notebook dangling from the cord around her neck.

Do NOT panic, Lucy, it read. *Enter kitchen for breakfast. Make tea. Eat cereal with milk. Eat fruit in fridge.*

That was a lot to remember, Lucy thought. There was a rich smell of cat litter and sardines out in the hall. She should probably empty the tray, she thought.

A white sticker was on the door. *Kitchen*, it read. She pushed open the door.

A tall, cadaverous man in a pale-blue short-sleeved top stood in her kitchen, by her pantry door. He sneezed with some violence as a cat wound itself round his legs. Another cat sat on her windowsill, apparently watching the man. As he turned at the noise of the door, he frowned at Lucy, his eyes watering and his nose red. He gave an irritated scratchy cough. Then, as if remembering himself, he attempted a smile. *Wary*, the word came to her. As if expecting she might scream or run away. She knew not to do that. She was known to him, she could tell. Which meant she should know his name. It was so rude of her not to know his name.

'Hello again, Dr Sampson,' he said, his accent American, his tone nasal. 'Long time no see.' He rubbed at his long thin nose with the back of his hand as if it troubled him. He needed a tissue, she thought. Did she have any?

Lucy found herself staring at the kitchen table. She must use it as a workstation, she thought. An ancient computer stood on the table. Two large files next to it, along with a vase of dying flowers. Cupboard doors stood open and two drawers had been pulled out as if the man had been searching for something. Or had she left it that way last night? Had she been searching for something? She often was these days.

She had learned not to protest at the presence of strangers in her house. *Carers*, the word came to her. This startlingly thin man is a 'carer'. Although Lucy did not think he had the face of a carer. She didn't like his face at all. Such high and bony cheekbones, as if he was starving. Should she offer this 'carer' something to eat?

The man's right hand slipped into the pocket of his blue top.

'Lucy,' the man said. He had something in his hand, she thought, although she couldn't make out what it was. 'I thought we might run you a little bath.' He had such cold, black eyes, she thought. 'Would you like that? Don't fall asleep in it though, will you?' He gave another scratchy cough.

Instinctively, Lucy took a step backwards. 'I have to feed the cats,' she said. Her eyes went to his hand. It was a hypodermic needle. Was she diabetic now? That really was too bad.

She opened her mouth to ask him and as she did, the front door opened, the yellow, pink and orange Post-it notes that covered the walls fluttering in the breeze.

'Wotcha, Luce,' Ash called out. She had a cigarette in her hand; a plume of smoke escaped her lips and she flicked the cigarette out of the front door onto the garden path. 'You're up and dressed. That's a good girl. Thought you wouldn't mind if I came at lunchtime today. Half the team are off sick and I've been run off my feet, like you wouldn't believe.'

Lucy looked back into the kitchen in confusion. Wasn't there a carer here already? But there was no one.

'Bleeding hell, Luce. It's perishing in here,' Ash said, stepping around her and making for the back door, which stood wide open. 'Let's shut this for a start, shall we? Fancy a cuppa?'

24

Murdo

The island was larger than North was expecting, with jagged teeth of rocks along the westerly side, brooding cliffs and a huge bay, at the centre of which he could make out a quayside. Small stone cottages were built into the hillside, some of them apparently in ruins, and, ranked further back, what looked like overturned boats, but which were obviously some kind of eco living quarters. A huge concrete and glass building stood to the west and, along and down from it, something low-slung in a dark aluminium. A sprawling stone-built manse, painted white, stood above it all as if keeping watch. Even though the settlement was a mix of old and new, and the architecture of the new as modern as he'd ever seen, viewed as a whole the place had a distinct and sinister harmony.

He wondered how it would have struck the Vikings and traders a thousand years ago. Whether they were the ones to first call it cursed. He felt lonely just looking at it.

'Beautiful, isn't it.' Comfort turned with a radiant smile.

North nodded, but it was a lie. Although the green and bronze land set against the immense silver sky and the silver seas was beautiful. Even with the rise and fall of thousands of seabirds, there was something stark and forbidding about the place. Perhaps it was the fall of steep jagged rocks or its apparent barrenness, but it made him think of islands used as prisons – places of exile and separation. It made him think some places were never meant for men to settle.

'No trees?' he said, suddenly realizing the lack of them.

Comfort gave a small laugh. 'They don't grow on Murdo. You get used to it.'

Then, as if she'd read his statement as a criticism, she spoke again, her tone defensive. 'We do great work here.' Her eyes were fixed on the island. 'Our community will open the yett between here and the next simulation, between this universe and others.' She seemed entirely sincere to North, her face a picture of shy pride. North wondered if she would feel the same way if she knew the bloodshed Enoch was planning. He glanced down at her. Could she know? She was barely twenty, with the face of an angel. The rose-tattoo girl, maybe. But this girl? Not for a second, he thought.

She had to know Enoch, though. The kind of man he was. He'd have asked about him then, but the engine started growling in protest and the smell of diesel filled his nose as the captain manoeuvred the boat against the stone quayside, the sea knocking the tyres that lined its sides against the concrete. The Zodiacs came in alongside, throwing out their own smells and noises, their passengers still yelling and whooping, and North realized that along the quayside there was the lift and urgent call and thrumming beat of music played by fiddles and bodhrán. It sounded like the band was out to celebrate the boat's safe passage. Joy flung a rope out to a weather-beaten man with the appearance of someone who'd been watching for them all.

She stepped lightly from the boat to the quayside and stumbled as she attempted to right herself. North thought she wouldn't be going on any more sea voyages for the foreseeable future. Comfort followed her, before gesturing to North.

North hesitated, reluctant suddenly to leave Mac and the boat, the wailing reel louder now, the drum's tattoo like thunder coming on. 'How long will you stay?' he said to the captain, who shrugged as he looked to the still-forbidding sky.

'Long enough to do the necessary to my engine, get some dry clothes on and something hot in my belly, and then I'll head back home.'

'Shouldn't you rest up? You were in the water a fair time.'

The captain looked offended. As if almost drowning was the worst excuse he had ever come across for resting.

'Nope. And remember, lad. These aren't good people. I don't ever stay longer than I have to and you shouldn't either.'

He turned away and called out to the weather-beaten man to bring him a change of clothes and some hot food. Joy whispered something to the man, who trudged away. North thought he hadn't much appreciated being ordered around by Mac. Over her shoulder, she threw North an angry look as if now they were safe back on the island, she resented his presence. He gave her a sombre nod as if to remind her that the Universe wanted him here, before swinging his rucksack onto his back. He was getting off the boat and everything would be fine. Meant to be. Joy took the hint.

Mac had the right idea about one thing. Getting changed. North's own wet clothes rubbed against his skin and his boots squeaked as he climbed up onto the sill of the boat and onto the side. The hard stone felt strange under his feet. He was on dry land, he told himself with relief, yet his traitorous body hadn't yet given up the sway and churn of the sea. Comfort reached for his hand and slipped hers into it. The girl held on to him like they were teenagers on a date, he thought. North turned his wrist ever so slightly to break her hold, but it merely made her cling to him more tightly.

If he was prey, then she wasn't losing him at this point.

Or perhaps she needed him there?

As he walked up the quayside away from the boat and into the face of a stiff wind, North turned back and raised his free hand. The captain was watching him go, a frown on his face. Mac made no gesture in return as if North was no one, as if he hadn't saved his life, as if he should have listened to his warnings. North stopped walking, feeling a sudden desire to return to the boat. Again, the memory of Jonny Gaffney's staring eyes, the swinging bag around his neck, and the endless drop into the Thames.

'What about your supplies?' North said to Comfort, thinking about the crates of alcohol in the hold. He felt as if the wind was trying to lift him from the quay and dump him back in the sea. She started as if guilty, then shrugged. 'Brothers and sisters will come for them. And I'll make sure they bring Mac down some hot food and coffee. He knows he can stay and rest up if he wants to, but he won't.'

The music had stopped, he realized. Now the only sound was the wind and the sea.

Comfort looked flaked out all of a sudden, the shadows rendering the pretty girl lifeless and flat, her hair like weeds. 'I don't care if I get into trouble for bringing you back with me,' she said, her voice dropping to a low whisper. 'I'm just so happy you're here.' She stood on her tiptoes to kiss his cheek. He had been so focused on getting on to the island, he hadn't considered the immediate consequences for Comfort and for Joy. That was wrong of him.

He had a job to do – shut down the broadcast, shut down Enoch and take apart a cult. Surely it was better for Mia and for Comfort if he was here? Providing he was successful, he could even help Comfort leave this place and rejoin the world. She was young. She deserved a shot at a normal life, didn't she? He couldn't afford to fail for all kinds of reasons. Mia. Thousands of innocent people who might otherwise get hurt. And now Comfort.

As if she sensed that she was on his mind, the girl tugged on his hand, and he realized he was either going to have to start walking again or turn on his heel and return to Mac.

Movement ahead of them caught his eye. Joy was moving fast towards a group of islanders swirling around something on the ground. There was a sudden scream.

25

London

It was one o'clock by the time Fang got back to her suite at the Empire.

Across the room, the giant TV was on, the noise of a soap opera blaring from it. Oblivious, Granny Po napped on the brocade couch in front of it, her small socked feet pointing upright, an empty noodle bowl on the coffee table next to her. Fang slid from her seat and padded across the room to scoop up a rug. The old lady slept so silently there was always a chance she was dead, but Fang knew better than to poke her or hold a finger in front of her nose. She had done it before and the old lady had a habit of opening her eyes and seizing hold of her granddaughter's wrist in a vice-like grip. Haranguing her in Cantonese for some considerable time along the lines that she wasn't dead yet and that she would know when she was dead, because she'd haunt her. Fang figured there'd been times when her gran needed to be vigilant in her sleep to stay alive, and it showed.

She laid the rug across the old lady's legs and Granny Po, with her eyes still closed, patted her granddaughter's hand in appreciation. The touch warm and soft and achingly familiar. As unlike the touch of the stranger on the bus as it was possible to be.

As quietly as she could, Fang moved away from the sofa and crossed over to the kitchenette. She took a can of Diet Coke out of the fridge, dropped way too much ice in it, and, holding the tumbler against her cheekbone, brought it back to her desk. She had tried to get away with hiding the swollen cheekbone from her grandmother by rearranging a sweep of hair across it. Fang had no compunction about lying, but in her experience lying to her grandmother was impossible. She was safe only as long as Granny Po's eyes were shut.

Fang didn't want to get shipped across to some detention centre in the US and be forgotten. It was a prospect that brought her out in goosebumps and left her nauseous. But she wasn't merely frightened, she was angry, at her own stupidity in falling for the chatroom con, and at that stuck-up spook with the milk teeth who thought she had her under control. Fang refused to be told what to do by the American. Or anyone else – aside from Granny Po.

On the one hand, her freedom was at stake. And indeed Granny Po's life. Because she feared Granny Po would wither up and die without her. On the other, there was the small matter of her moral code. If she did what Roxie wanted, she'd have to give up Hone and her country, even if she managed to keep North clear of the fallout and herself out of prison.

It was up to her to figure a way out of this mess. As she saw it, she had three immediate tasks.

The first: to brief North on what she and Plug had discovered about the Narrow Yett's charitable giving and in particular Craig Drummond. Drummond was a nasty piece of work and he'd get even nastier if he thought North would be a threat to his money-laundering operation. North needed all the help he could get if he was to stop whatever was supposed to be happening in the next – she checked the clock – twenty-two hours and thirty minutes.

The second: to figure out exactly what Benson Corporation was up to on Murdo with their commercial spaceport and what had earned them special treatment. In particular, what was the mystery building Bald Paulie had pointed out? Because the moment Roxie punched her had made the Benson Corporation, and all who sailed in it, Fangfang Yu's absolute top priority.

The third: to get herself out from under Roxie, before the agent got to put her on a plane to Nowheresville, USA. And part and parcel of that was to keep North out of the mess she'd gotten herself into. To keep him focused on the Narrow Yett because if she so much as hinted that Roxie was threatening to fly her across to the US with a black bag over her head, he'd be back in London before she could say 'moron-person'. But Fangfang Yu didn't need North to rescue her. And she didn't need North getting himself dead. Until she had the full picture on the spaceport, she was keeping him out of it. He had enough on avoiding castration, stopping the broadcast, dealing with a bloodthirsty cult leader, and keeping clear of a violent ganglord.

Even as she was thinking, her phone pinged.

'Good talk today, Fang.' It was Roxie. 'I'll drop by later and you can tell me what your decision is on the small matter of treason. Don't disappoint me. I'm not good with disappointment.'

It was a new phone. Should have been untraceable. Roxie wanted to let her know there was no such thing.

But whether she was good with disappointment or not, if Roxie thought Fang was going to betray North, she was about to be disappointed. Fang considered how Roxie dealt with disappointment and her fingertips went to her swollen cheekbone. Roxie was a bully and Fang had met with enough bullies to know how they operated. For a brief second, she allowed herself the luxury of considering Roxie's death. Fang had access to a great deal of money. She'd also made it her business to keep track of an assassin-for-hire known as Lilith. Not because she'd ever considered she would need her services, but because she didn't trust Lilith to maintain her truce with North. Lilith, she knew, would have no qualms about taking Roxie out if Fang paid her enough, even if she was a US agent. But Fang wasn't comfortable with violence, unless it proved strictly necessary. Plus, if she was being honest with herself, however obnoxious Roxie was – and Roxie was deeply obnoxious – the agent shouldn't die because she was simply doing her job. More importantly, Lilith would love the idea that Fang needed something from her and Fang hated Lilith's guts even more than she despised Roxie, which was saying something.

She undid one plait and then the other, before plaiting the silky black hair up again. She pulled across her laptop and allowed her fingers to graze the keys of the keyboard, closed her eyes then started typing. At first, her focus was splintered – Roxie's tiny hand, the sensation of her own face slamming against the glass, Bald Paulie's question, *Are you in trouble?* But eventually, the familiarity of the movement soothed her. The code unfurled as she fashioned her own path – her speed building all the time – through the highways and byways of the net, the fizz and pop of blue light, transporting herself through commands and brackets and spiralling numbers in her urgent quest for information about the Benson Corporation.

What was the company up to on Murdo? It had to be more than setting up a spaceport to deliver communications satellites into low Earth orbit. That was all a matter of public record. Nothing that would prompt the involvement of the agency employing Roxie Powell. As Fang saw it, either there was something odd about the nature of the satellites going up or rocket launches and satellite delivery weren't the only thing going on there. And if there was something else, what were the chances that it was going on in the mystery building?

But there was nothing she could see that made sense or that tied the Benson Corp to anything untoward. She pulled up the company's staff files and started working through them, then sighed.

This was a big piece of work. It was a global company with thousands of employees. Even if she narrowed it down to the space division, she was looking at hundreds of people. Before she got into it, she needed to be able to work without worrying that Roxie would knock at the door. And when she said knock, she meant knock the door down and then stamp on her head. Fang needed to disappear for a while and Granny Po with her.

But Roxie was in a state of hypervigilance.

Roxie had said she was going to 'drop by'. But Fang needed to buy herself some time.

Fang had propped the business card with Roxie's mobile number up against the fruit bowl the hotel kept stocked with exotic fruits. A dozen tiny fruit flies hovered over it. Wrinkling her nose in distaste, she worked her way through the maze of consequences, thinking three moves ahead, twelve moves, one hundred. Would North come to any harm if she told the American what she had found out about the Narrow Yett's money? About Drummond's criminal activities? She didn't see how, and if she appeared to be cowed and willing to the agent, that would at least give her the breathing space she needed to keep working on Benson without Roxie being any the wiser.

Then again, if Hone found out Fang had given up secrets to the Americans, there was no saying what he was capable of. He'd be as keen to throw her in a UK jail as Roxie was to throw her in a US jail. Hone had never in her experience shown himself worthy of trust. He'd once drugged and kidnapped her to force her to work for him. Only a couple of months ago, he'd snatched her mother and thrown Mama Yu in a detention centre to make Fang work for him again, when he could have just asked. There was nothing the man wouldn't do in the name of national security. Having a traitor on his team would be met with extreme prejudice.

She was in this on her own. And sure, part of her wished she could run her play past North. But she couldn't. Even so, she wanted to hear his voice. She checked the GPS for the dot that showed his location. He had arrived – finally.

26

Murdo

North started running. He wasn't sure why. An instinct that told him someone was in trouble. He shook off Comfort. Heard a whimper, felt her drag against him as she tried to keep tight hold of him, but this time he wasn't letting it happen.

Dozens of people in charcoal-grey joggers and fleeces jostled each other as they streamed after a huge man barrelling down a cobbled lane towards a young boy, hampered by the weight of a small girl, her hair in pigtails – and running full tilt towards the quayside. As he watched, the child wrapped plump arms around the boy's neck as he switched her weight from one hip to the other, and started wailing. Was the boy going to throw the child into the sea? It didn't look like it from the way he was holding her close. But there was nowhere else for him to go.

Even as North ran, something about the lumbering gait of the boy's pursuer put him in mind of a bear. North thought of what it would be like to be pursued by a raging bear. Terrifying, he was guessing, his gaze fixed on the white-faced boy coming full tilt towards him. The boy glanced behind, and as he did so the toe of his trainer caught on a crack in the stone and he stumbled. The bear-man was carrying too much weight, but even so, he was fast. A huge hairy arm reached for the boy's shoulder, swinging him around and almost off his feet. The toddler let out a shriek as she felt the boy holding her lose his balance, but the man kept them both upright.

Roaring, the bear-man seized hold of the child, ripping her from the boy's grasp, and in the same motion used the flat of his hand against the boy's chest to push him full force away from them. The boy staggered, his lean form tipping over, toppling, a cry of protest caught in his throat. And as he disappeared over the edge of the quayside, North reached for him, grasping the skinny arm, hauling him back on to land.

The boy gave the stranger a look of confusion rather than thanks, as if he hadn't expected anyone to save him. It took him less than a second to find his feet and less again before he moved forward in an attempt to get hold of the screaming child the man had already handed off to someone in the crowd.

'Rowan, you little scumbag,' the bear-man yelled, spittle flying in all directions. 'Where do you think you're taking her!' The big man pushed the lad off and away, but this time the boy managed to stay on his feet. It wasn't the first time he had been on the receiving end of violence from the man, North thought, and he knew from his own experience that Rowan would have a feel for it – know the pattern and choreography of it. The odds against him were overwhelming but he'd feint and dodge, and do what damage he could, if the man was predictable.

North had come across his share of intimidating characters in his time. In his line of work, it was hard not to. He knew the difference between the blusterers and the dangerous. Some men liked to talk more than they liked to act, and some men liked to do violence a lot more than they liked to breathe in and out. The bear-man appeared to be one of the latter. North felt bad for the boy, even if he didn't know the rights and wrongs of what was playing out in front of him. He took a few seconds to scan the faces of the crowd and caught the fear and the judgement on them. They were frightened of the man's temper and violence, but even so they judged the boy to have done something that merited it. No one seemed willing to speak up for the lad, which sat badly with North.

The lad came at the big man again. Not to get purchase on the child this time, because she was sobbing now in the arms of a whey-faced woman he was guessing had to be her mother. No, this time Rowan came with fists flying. He was rangy, with long arms and legs, but no older than thirteen. Just beginning to show who and what he'd be as a man, but he had no strength behind his blows. The man's brows gathered in a thick line and his knuckles tightened. So far, he had batted the lad off as much to humiliate him as to hurt him, North knew. But this blow would be different. This was intended to hammer the boy into the ground, to hurt him and do what damage could be done. The bear-man pulled his fist back and made ready, before he started driving it forward.

North stepped forward and caught the man's thick wrist in his steely grip. 'You don't want to do that,' he said.

The man growled at the back of his throat, and it was the noise an animal makes before it attacks. 'Are you telling me what to do with what's mine? I wouldn't do that.' The teeth were bared and very white, and close up North saw the hair was close cropped, the beard oiled and kempt.

'I'm not yours,' the boy yelled.

But the bear-man had lost interest in the boy. He waved him off in the same motion as he turned his body square on to North's. The eyes were cold now, the temper not so much in evidence, making him more threatening rather than less. This was a powerful man on his own territory with friends behind him confronting a stranger who had no right to be here and no friends to be seen.

'And who the fuck are you?'

'I was passing. Thought I'd drop by.' North kept his voice pleasant, polite, but he watched the other man's fists, watched his eyes go to the crowd. He couldn't back down. There were too many witnesses. Instead, the big man signalled to a couple of heavies, who crossed to the lad and took hold of him. The boy squirmed in their grasp but they held on too tight for him to break away. The bear-man gestured towards Mac's boat and they started pulling him across. Rowan struggled and started yelling, 'I'm going nowhere. Let me go. I'm staying here. Mam! Mam!'

His mother made no move towards him. Her expression was dispassionate, North thought. If she had ever cared for the boy, there was no glimmer of it there now. Rowan was an embarrassment – nothing more. Instead, she turned, holding her daughter closer, and pushed through the crowd, her back to her son as she made her way up the cobbled street.

North blocked the path of the two men dragging the yelling boy. 'He doesn't seem to want to go.' The heavies looked back towards the bear-man. 'Drummond?' one of them said, awaiting instruction. Drummond was someone round here, North figured. Used to having his orders followed.

Behind him, he heard Mac call out, 'I'm taking no one across without they want to go.'

Drummond moved closer to North, each step heavy with the threat of violence, and North sensed the spectators shrink backwards as if they had seen him walk that way before and it never ended well. He was so close now that North could feel his warm breath on his face. Too close to throw a decent punch, he thought. This was about intimidation first and then violence. With most men, the intimidation would be enough. But then North wasn't most men. The recognition of that fact sparked in Drummond's face, and he drilled his index finger into North's chest.

'You'd be advised to get back on that boat while you still can, pal. This is private property.'

North kept his face relaxed, although his body was tensed and ready for action. 'The thing is, I'm queasy from the trip. I thought I'd stay and find my land legs, you know.'

Drummond took a step back, his arm swinging. North's arm was swinging with it to block the punch when he heard the woman's voice.

'Hey, Father of the Year! If you're going to hit someone, give me a minute. I need to frame the shot.' By a pile of lobster pots, a young woman in a bright-red parka with a fur-trimmed hood was lifting a camera onto her shoulder.

Mia. He didn't have to see her face. The smoky voice took him straight back to childhood. The smoky voice and the fact she was figuring to protect him and the young boy. Mia Barton really was here on Murdo. And he was here with her. After all these years.

Drummond turned towards her at the shout. His guard down. North didn't know what he expected. Probably not a punch to the head. But as far as he was concerned, Drummond started it so he would finish it. Mia and he had played this game before. He'd get into a fight. She'd provide a diversion. And he'd take his opponent any way he could. The big man staggered then went down, as North expected, his body sprawled over the concrete, his head slamming into the ground. In unison, the heavies let go their hold on Rowan and hurried over to Drummond. North glanced at Rowan, his eyebrows raised, and the boy took the hint. Sparing North a grin, he took to his heels, running backwards for a while as if he couldn't get enough of the sight of Drummond laid out on the quayside. With a loud whoop, he waved to Mia, and then took off across the scrublands. The youngster could certainly run. But if Drummond was his mother's partner, which was North's bet, that was understandable.

Drummond was roaring, but his friends were holding him back. He was furious but seemed dazed as he attempted to push them away. Whatever his capacity for violence, they had obviously decided that it had to be hampered by getting knocked off his feet. He didn't seem to share their opinion. But he wasn't beyond reason. They were pointing at Mia, who was filming it all. The camera still to her eye, she waved with one arm as if to say: *Don't mind me – carry on.* And Drummond appeared to remember why he'd stayed his own punch.

North realized that Mia had the clearest view possible of him. Even if she had been playing peacemaker and hadn't realized it was him at first. She knew now.

'You—' Drummond pointed his finger again, keeping his voice low so the camera couldn't pick it up. '—pal, better watch your back.' And with that he turned, shaking off the supportive arms of his two mates. 'And *you* do not have my permission to film me. Or my men. Ever.' With a final glower at North, he lumbered away. Rowan, it appeared, was forgotten.

'Poor kid. I can't stand that guy.' Comfort was suddenly alongside. Her eyes on Drummond's retreating back, a look of distaste on her face. 'But you've got a regular saviour complex, huh?' She gave his arm a small squeeze before she crossed to the spectators and started pointing them back along the pier. He was guessing she was asking them to help unload the supplies from Mac's boat.

Mia lowered the camera and turned her back a little to keep the wind from her. She appeared to be checking the footage.

He kept his eyes fixed on her. She brought down the hood. Dark ringlets squirrelled out from a white woollen beanie. From the angle where he stood, he still couldn't see her face, only a glimpse of her side profile, but he knew from past experience that if she was focusing hard on something, her right incisor would have caught her lower lip and her brows would be gathered together in a look of intense concentration. He also knew the skin would be a pattern of white and brown from the vitiligo she had, white around the mouth and the nose, around her eyes and over most of her hands. She'd been teased in the children's home about it. He'd told her she was beautiful, which was the simple truth.

She glanced across at North over her shoulder as if to check something against the footage, and he felt the recognition in the warm brown eyes like a jolt of electricity. He realized he was beaming and stopped. His instinct was to go to her, to pick her up and swing her off her feet, as he'd done when they were kids. But Comfort would be sure to see – it would blow his cover and the operation. Instead, he gave the slightest shake of the head. A warning, that anyone watching would read as disapproval that she'd been filming.

'That was quite some arrival,' she called over to him, snapping off the camera and moving off. 'Well worth the wait.' She didn't look back at him, just moved away up the quayside and towards the settlement as if she had places to be. And as he watched her go, he wanted above all things to call her back. 'Well worth the wait'? To anyone listening, she would have been referring to the fact the journey across had taken longer than it was supposed to. But he knew it was more than that.

'She's going to make us famous.' Comfort had slid in beside him without him realizing. Her tone was wistful, as if fame was something she could only dream of but already knew it was for other people and not meant for the likes of her.

'Who is she?' North said, although he already knew. Had she recognized him? He thought so. Hadn't she? He began to doubt himself. But no, of course she had. He couldn't help but feel almost drowning had been worth every second for the sight of that face. Hearing that voice.

'Her name's Mia Barton. She's filming a reality TV programme. Enoch says it's part of our mission to reveal the truth about this world.' North said nothing. 'Drummond hates it. He's refused to be in it, won't sign any of the permissions, but he can't stop her filming everyone else. Mia and Enoch are tight, you know what I mean.'

Tight? Was she implying what he thought she was implying? He had to hope not.

He heard Mac call his name and, as Comfort waited, went to make his farewells. When he offered his hand, Mac kept hold. He turned so Comfort couldn't make out what he was saying. 'I don't know what your game is, laddie, but I figure you're up to something. Drummond and his pals are a bad lot, though, and you've made an enemy there.'

North rested his other hand on Mac's shoulder. 'Thanks for an eventful journey, Mac. Don't take it personally, but I might call an Uber next time.'

Mac stepped away, shaking his head. 'You're funny with it, huh. Well, Mr Laurel and Hardy, there's more bad weather coming and I won't be back out here for a while yet. You might not last that long with the likes of him around. And Lord knows that boy would be better off away from this place. I won't take him against his will but if you can change his mind, when I come back, I'll take you both off the island. Make sure you stay alive that long.'

North laughed but Mac was right to be concerned. Drummond gave every impression of a man who bore a grudge. As he turned away, he wondered about the wisdom of what he was doing. If Mia was indeed 'tight' with Enoch, she certainly wouldn't want North killing him. Could he even trust her with the intelligence that Enoch wanted violence and mayhem on the streets? Would she believe him if he did tell her what was going on?

And she was hardly in any kind of danger. Unless his arrival put her there. She'd spent five months embedded in the cult, doubtless building all kinds of relationships. But he wasn't like her – he wasn't on the island to bear witness, he was on the island to destroy Enoch and if necessary, everything Enoch had built out here. And he'd been here less than five minutes and he'd already made his first enemy. What was the betting it wouldn't be his last.

27

As he and Comfort slogged up the track into the teeth of the wind and towards the houses, he noticed a figure watching them from the upstairs window of a large stone house high above the settlement. This must be the manse, the paint white and peeling, and he'd put any amount of money that it was where Enoch lived. Was he watching the arrival of a stranger? Had he broken off from his plans to spread mayhem and bloodshed? Enoch, North thought to himself, was about to find out just how it felt to be on the receiving end of mayhem and bloodshed.

He squinted his eyes to try and make out the watcher, but as he blinked, the shape shifted and disappeared.

Comfort was talking to him. Something about seabirds. He zoned back in. She had stuck feathers in her blue-green hair, plucked from the ground, and as he watched they blew away on the breeze. 'I'll buy you some more,' he said, still trying to channel the patrician cover story. She giggled. Being charming was exhausting, he decided.

They weren't heading for the manse, which had been his first assumption. The houses were low and stone built. One in every three appeared to be derelict – their roofs caved in on themselves and the walls tumbled in – but most were in good repair, with smoke easing out of chimneys. Beyond them there were larger buildings, sprawled and modern. One with an immense picture window looking out onto the sea. The other constructed around an ancient church.

North sensed he and Comfort were being watched by the inhabitants of the settlement. Joy had made it clear that Enoch said no visitors on this run. But here he was anyway. Did the inhabitants disapprove? His fight with Drummond wouldn't have gone down well. Comfort was a sweet kid and he hoped that Drummond wouldn't blame her for what went down at the quayside.

The blue-haired girl hovered at the doorway to a cottage.

'What's up?' he said.

Comfort screwed up her pretty face. 'We have accommodation for visitors, but that's usually in summer. And anyway, Enoch said the celebration was only for family.' She meant members of the cult, he guessed. 'It's our tenth anniversary of when we first came here.'

Was Enoch planning to coincide his call to arms with the cult's anniversary? That would make sense.

'And I'm crashing the party?'

She shrugged. 'I really need to go sort things with Enoch before Drummond gets to him. Enoch's our leader, but Drummond's important – he runs the whole cyber operation of the church. I can square it with Enoch if I explain everything that went on with the storm. I just want you safe while I do that...'

'Safe?'

She looked pained. The huge blue eyes downcast. 'It's my fault. I should have stopped you getting involved. Drummond's a thug and you showed him up. He won't forget. Once Enoch gives the okay, though, Drummond will have to get on with it. Meanwhile, I don't want his guys bundling you up and taking you back to Mac's boat. I know this cottage is empty because...' Again, she hesitated. Had someone died here? Or quit the island? Could this place possible be Gaffney's?

He pushed open the door, as if intrigued to see what was beyond. There was a chance it was colder in the cottage than it was standing outside in the wind. Comfort followed him in and shivered. It smelled of damp and it was dark and full of shadows – exactly the kind of place a man inclined to madness would feel at home.

'Blood and sand, it's cold in here,' she said, crossing to the fire and rummaging around in the hearth. North let her get on with it while he considered his surroundings. The wind pushed up against the leaded glass, and he could feel the cut of the cold air as it eased its way in through the cracks and crevices between the frame and the stone. The place was austere but as far as he could make out in the gloom it was clean. Had Gaffney left it this way? He didn't think so. Gaffney's mind had been a ruin and in North's experience that didn't lead to neatly made beds and wiped-down surfaces. Someone had been in here after Gaffney left and cleaned house. A crackle as Comfort got the kindling alight and added sticks. No waste. Blowing on the feeble flame, it caught, and when she was satisfied she added a few logs of peat from the rusting bucket set close by the hearth.

As she stood her cheeks were flushed and she grinned, pleased at her own expertise.

'We cut the peat ourselves,' she said. 'There are peat bogs around by the loch.'

Then she stopped talking, the look on her face changing to surprise and then to concern. North followed her gaze. The fire had chased the shadows from the stone walls where Gaffney had made his mark. He had to have drawn with charcoal, because black ash came away as North withdrew his finger from the plasterwork.

Gaffney might have been an accountant in a previous life but had perhaps missed his calling. As an artist, he was primitive but effective. To the right, naked men and women were drawn lying together in piles, all with closed eyes. More were pictured standing, their eyes wide open and their mouths screaming, some of the men clutching their members in their hands, black blood gushing from where they should have been attached. More still were clambering over each other in obscene displays of coupling. To the left, a huge phallus with flames coming out of its base, and in every clear space scrawled strings of computer code. It made for uncomfortable viewing, not least because the whole scene was framed in what looked like bloody handprints interspersed with writhing snakes.

The firelight cast flickering shadows that made it look as if the hellish figures danced and shook and copulated. Comfort was quiet as she stood close by him scanning the artwork as if unsure as to what to say. 'Sorry about this. The guy who used to live here went bat-shit crazy. I can't believe it hasn't been painted over already, I told you we weren't expecting visitors.'

No kidding.

'Did you know him?'

'Of course. He was a nice guy. At least, he used to be.'

'What do you mean?'

Would she mention the castration, he wondered?

She stood still, as if considering her response required all her concentration and energy. 'Before I came here, I worked in a zoo. It was truly terrible. I hated it, but it got me into animal rights. You know how animals can go crazy in a zoo? They call it zoochosis.'

Do they? North had no idea.

'They hurt themselves and they do the same things over and over. They're traumatized by their captivity.' She took a deep breath. 'I think that's what happened to Jonny. And he wasn't the only one.' Her face was pale when she looked at him. 'Not everyone can deal with the fact we exist in a simulation. It takes discipline to live within this reality and not go stone-cold crazy. Put that together with the fact that some people are loners and shouldn't live in a community. Not to mention living on an island. His system blew up.

'A year and a half ago, we had a party. It wasn't planned, but it was a wild night. I can't remember any of it, really. But the morning after, we woke up to find Jonny screaming and running around naked. He'd gotten hold of a knife...' She shuddered. 'You have no idea. We've a doctor here and a health clinic and he recovered. But he was never the same.'

That didn't surprise North.

'He suffered from anxiety – more than that, he was paranoid. It was a difficult period. Jonny wasn't the only man to mutilate himself. A few of the guys became adamant they had to do the same thing. Enoch brought in a therapist, but nothing worked. Eventually he said who were we to judge, but we should make sure everything was done in hygienic conditions rather than have them butcher themselves like Jonny. Two people even killed themselves. Really sweet people. A husband and wife held hands and leaped from the cliffs into the sea. A handful of people went back to their lives on the mainland. They got too anxious to stay. For a couple of months, things went off the rails here.'

North found himself more disturbed than he had expected. Mac had said he'd heard stories. But this was the reality. Gaffney was ill.

Comfort reached out a hand and laid her palm against

one of the bloody handprints, as if trying to ease Gaffney's suffering. 'But everything calmed down. It went back to how it had always been, and Jonny at least seemed better. Then—' She pointed at the flaming phallus. '—you see the rocket.' Not a phallus then. 'You met the Americans in the bar. Tossers – as you saw by that trick in their helicopter. They sent up a rocket about six weeks ago. Jonny was out birdwatching. It was his thing. He just loved birds – he used to say that watching them tuned him into nature. But when he saw the rocket take off, it flipped some kind of switch in him again. He became obsessed with the fact billionaires like Elon Musk and Jeff Bezos were abandoning Earth for an alternative simulation because they know that the end of the world is coming any day now. Someone or something is going to press "quit".'

North was trying to keep up with her logic.

'Jonny was convinced that key players left on that rocket. Royalty. Billionaires. Politicians. He started pushing Enoch really hard on whether the yett would open in time for the rest of us to escape. Challenging him in public to figure out the code for us to break out. Eventually, he was asked to leave.'

'He was "asked to leave"? Like Drummond "asked" that kid at the quayside?'

She shook her head fervently. 'No, of course not. Absolutely everyone thought it was for the best.'

'You didn't believe him about the rocket?'

She shrugged. 'All I know is I trust Enoch. The yett opened for him once. And I'm going to be standing right next to him when it opens again.'

'And Gaffney?' North asked, as if wanting a loose end tied up.

'Last week, someone went with him to the mainland and

took him to a clinic to get help. I wanted to take him myself.'
Her lower lip trembled. 'But Enoch said it would be too
distressing for me. At least now Jonny's getting the kind of
help he couldn't get here.'

Gaffney had needed help, but he didn't get it. Instead, he'd
hurt any number of people and then damn near decapitated
himself – facts Comfort appeared oblivious to. On tiptoes, she
reached up and he felt her cool lips against his cheek again,
her small hand on his shoulder. 'Are you wishing you hadn't
jumped on our boat now? I promise you we aren't all crazy.
It's really not like that any more or Enoch would never have
allowed Mia to come and film us.' She smiled. 'Settle in, and
as soon as I've squared things with people, I'll be back with
coffee and something to eat to keep you going. I'm really
happy you're here, North. Get some rest.' The door opened
and closed after her. She was right about his need to rest up,
but who had time for that?

What the hell?

He didn't think he'd ever feel sympathy for Gaffney after
what he did on Tower Bridge. But the guy sounded as if he'd
had the most catastrophic breakdown. And instead of being
someplace safe, he'd been unlucky enough to be on a rock in
an ocean surrounded by people who didn't know which way
was up. North shouldn't have jumped from Tower Bridge. He
should have saved Gaffney. The man needed help.

North had changed out of his saturated clothes on the
boat, knotting them into a bundle. He unknotted it now and
draped them over a wooden chair he pulled closer to the fire.
Then dumped the rucksack on the bed, unzipping the secret
compartment. He checked the gun had stayed dry and then
pulled out the satellite phone. He was touching base with Fang
and then he was scouting the island. No, he wasn't wishing he

hadn't come. He thought it was a damn good job he was here.

His more immediate problem, though, was making sure he kept out of the way till Mac's boat had left harbour. But first he needed to talk to Fang. He pressed a button on the phone and it crackled into life. Fang's voice cut through the crackle of the fire and the sound of the crashing waves. '*Come in moron-person.*'

28

'Yes, I got here fine, thanks for asking, Fang.' He thought it his duty to remind her of her manners. She ignored him.

'What do we know, Fang?'

'*The Narrow Yett isn't just a cult, it's a criminal organization.*' Was it his imagination or did she sound distracted? She was probably doing six things at once, as ever. '*I'll keep this simple for you because I don't want to make your head squeak, so let's just say I believe the cult is laundering money through its charitable giving.*'

Money, North thought. It never brought out the best in people.

'*But it's worse than that,*' Fang said.

Worse than a cult?

Worse than a cult that laundered money?

'*One of the charities it's involved with is being used as a holding station for women being trafficked into sex work.*'

Yep, that was worse.

'*The guy in charge of the operation is called Craig Drummond.*'

Drummond. The bear-man on the quayside. The violent bully who'd been ready to beat a young boy. The thug he'd humiliated. The one who'd said he'd better watch his back.

'*Apparently, he knows Enoch from university. He was the*

best man at his wedding. He's an out-and-out gangster. Does the name mean anything to you?'

Strangely enough, it did. And somehow he didn't think Drummond believed he existed anyplace but the real world. Didn't think he existed for anyone or anything's benefit but his own.

'Plug's checked him out with a contact. He's dangerous. Avoid him.'

North said nothing.

'North?'

'I've met him. He runs the cyber side of things up here. So he'd certainly have the opportunity to launder money.'

'And?'

'He doesn't like me.'

'Are you kidding? You only just got there.' There was a note of accusation in her voice.

Did he need to tell Fang everything? He didn't think so. She'd worry. He kept it to the minimum. 'Time enough to have a disagreement about how he treats his stepson. A boy called Rowan. When I stepped off the boat, Drummond was giving him a hard time and I objected.'

Fang let it go.

'Repeat: Drummond is dangerous. Do you get that? From the sign-offs on the accounts, Jonathan Gaffney acted as the cult's financial director. He had to be working with Drummond. Maybe voluntarily. Maybe Drummond forced him. But there's too much money sloshing around for Gaffney not to have known. At the hospital, Hone mentioned the Inland Revenue had an investigation ongoing. Maybe with an Inland Revenue investigation on the cards, it all got too much and it tipped Gaffney over the edge?'

Gaffney had abandoned his family, sold his house in

Edinburgh and given the proceeds to the Narrow Yett. He was a believer. Drummond must have forced him to cooperate. And the walls of his cottage spoke of a mind in turmoil rather than that of a criminal mastermind. Gaffney had broken under pressure.

He snapped a picture of the wall and pinged it to Fang. 'Want to see what over the edge looks like?'

There was a pause as she opened the file. '*Whoa!*' she said. '*He was not a happy bunny.*'

'Eighteen months ago, he wasn't the only one. Apparently, Gaffney had some sort of psychotic episode when he hurt himself, and that seemed to trigger general hysteria here. Other men decided castration was the way to go too. There were suicides and some members of the church up and quit.'

'*In the Middle Ages, you'd get outbreaks of mass hysteria occur in tight social groups, often groups united in spiritual beliefs. People would dance till they dropped. These days it's called "collective obsessional behaviour". Psychological distress triggers real physiological symptoms. It can be anxiety hysteria or the kind that affects motor function. It can happen. In the sixties, girls at a school in Tanzania were hit by a laughing epidemic. Teenage girls in New York all started showing symptoms like Tourette's, courtesy of some social media video.*' How did she know these things? '*It sounds like the Narrow Yett was hit by a bad episode.*

'*I don't like it, North. What's to say it won't happen again? Especially when they find out what their former pal did on Tower Bridge.*'

She had a point, but North couldn't speak to 'What if', so he didn't. 'Go back to the money for a second. Can you see who's holding the bag of money when the music stops?' he said. 'Is it Drummond or Enoch or both?'

There was the sound of Fang on the keyboard, the wet chewing of blue bubblegum he knew smelled of raspberries. *'No, that takes more time to unravel. But if I kept going, I think we'd also find any construction out there, any supplies and services the cult buy, they will be paying over the odds to companies that track back to Drummond. It's a regular cash machine.'*

North thought about it. This was a cult living on an island in the middle of nowhere, refusing to accept the reality of reality. A cult whose members had already suffered from an outbreak of collective hysteria. From what he'd heard so far, the cult members weren't a danger to anyone but themselves.

Of course, Drummond at least was involved in serious crimes on the mainland, including sex trafficking. Drummond was also using the Narrow Yett to launder money and commit fraud on a massive scale, with or without Enoch's cooperation.

But as criminal and reprehensible as all of that was, it didn't speak to Drummond's involvement in the kind of extremism that posed a risk to national security expected to kick off in – he checked his watch – twenty-two hours.

'Have you had time to check out those guys in the bar?' He kept his voice casual – stopping himself just in time from saying 'in the helicopter', as that would mean revealing it had flown over him when he was in the sea, busy drowning.

'Fang...?' he said.

'They're with a company called D'Urberville, who are a communications satellites business. D'Urberville is part of Benson Corporation. Benson Corporation is a commercial defence company with close links to the US government. You name it and they're into it. In fact, the Pentagon is their biggest customer.'

'The Pentagon?'

'Yep. And the US Space Force is involved somehow with that spaceport. Their Technology and Innovation Office signed off on the plans for its construction including three rocket launch pads.'

North had the impression there was more to be said, but again there was silence. He put it down to a delay on the line. 'And?'

He wasn't sure whether it was the distance, but he read reluctance in her voice.

'They're planning to use the spaceport to launch communications satellites up into the polar orbit. Low Earth orbit rather than deep space. It's big business these days. There is one thing that's odd, though. I met with Paulie and we checked the plans against a satellite image of the station. There's a building that stands apart from the others and we don't know what it's for.'

'Is that significant?'

'Maybe and maybe not. Maybe they thought no one would notice or care. Maybe they're getting on each other's nerves and wanted more space. Maybe they built themselves a spa.'

North thought about Kincaid and his meatheads. Fang was joking, he knew. But somehow he didn't think it was a spa.

'I'll go out and take a look. Comfort just told me some of it...'

'Comfort?'

'I travelled over with her on the boat. Apparently, there's already been a rocket launch. Gaffney saw it and it triggered him again.' North glanced at his watch. He had to wind up the call and get out of there. He really didn't want Comfort returning to tell him Drummond was insisting he got back on Mac's boat.

'So what's your plan? Are you just going to kill Enoch and

hope the broadcast doesn't happen because the rest of them jump into the sea like a load of lemmings? Shame they aren't still hysterical or that might work.'

What was his plan?

Good question.

'I'm going to recce the island and figure out who these people are first,' he said. 'I'm going to look for vulnerabilities and opportunities.' Attempting to sound more confident than he felt. What was his plan? He was going to make it up as he went. Like always.

'*If you want to make sure the broadcast doesn't happen, destroy their server and the satellite dish. If you take the server out, they won't be able to access any of the programs or data they need to keep their website and channels live. If you take the satellite dish out, they can't just make a Facebook Live announcement.'*

That also seemed like a plan.

'*And how's your head?'* Fang asked. Her tone was cool. As if the state of his head was on him since he'd flouted Linklater's advice and ignored her pleas. How was it? If he ignored the crashing headache and the pounding in his ears, it was completely fine. He told her so.

Afterwards, he stood with the gun in one hand and the phone in the other. He was tempted to keep both with him. Comms and a weapon were necessities of his trade. But there would be too many questions if they were discovered on him. With a sigh, he tucked them both back into the zippered compartment. Then left the rucksack unbuckled and propped up against a wall. He pulled out a small towel and draped it over the back of a chair, along with the jeans and hoodie, socks and underwear, all drenched with seawater. The damp T-shirt he left hanging half in and half out of the opening. There was virtue in leaving things allegedly in open sight. It was a double bluff. He had nothing to hide. Anyone who wanted to search it more thoroughly was welcome.

He slipped through the back door in the lean-to attached to the rear of the cottage, and took off across the hill, the wind pushing him backwards now as if it didn't want him to go. He didn't know how long he had, but he wanted to get the lie of the land. And if he happened upon Kincaid while he was there, he'd be sure and make an appropriate thank you for the helicopter's fly-by.

29

London

Roxie was perched on a bar stool. In front of her was a small glass of clear liquid, a lime and a salt cellar. She was staring at the glass as if she could imagine both nothing better and nothing worse than to drink it.

Fang perched on the stool next to her. The barman winked as he tipped ice into a glass, snapped off the lid of a bottle of Diet Pepsi and started pouring. He slid it in front of her, stepped back and tossed a maraschino cherry into it. It was his thing. Fang didn't like to judge.

'Did you notice my hand?' Roxie flexed the doll-like fingers.

What was the correct answer to a question like that?

'I noticed you hit me with the other one,' Fang said.

Roxie ignored the dig. 'The agency loves it. My hand ticks a lot of boxes for them on diversity and inclusion. Or it did, when all I wanted to be was a cyber threat analyst.'

Fang had no idea where this woman was going. Nowhere she wanted to follow, but it didn't look like she had a choice.

'It's a different ball game when you want to be a field agent though. There's a physical. I came first in the run. I can do sit-ups all day but the press-ups? No sweet way. Do you want to know how I passed that physical, Fang?' Roxie's eyes met Fang's in the mirror. 'I used my hand to take hold of the penis of the army instructor in overall charge and I said, "Look how big you are, soldier". He laughed like a hyena once he'd done staring. We're still on excellent terms.'

Fang took in three ice cubes along with a mouthful of Pepsi in the hope Roxie wasn't wanting a response to that.

'I'm telling you that so you realize I'm a big believer in figuring out where the problem is going to be and getting ahead of it. I've studied your work and it's almost flawless. Almost but not entirely, or you wouldn't be here. There is no move you can make I won't have anticipated. You need to know that.'

She went back to staring at the drink in front of her. 'So, am I celebrating?'

In that moment, Fang missed North more than she had ever missed anyone ever. Minutes ago, she'd almost caved and confessed she was in trouble. She'd said: 'How's your head?' and didn't even listen to his answer. Because what she had wanted to say was: 'Come home. I'm in trouble.'

If only she could explain what she was doing, but that was impossible. North had his own mission and he was hundreds of miles away. But even if he was here, standing right behind her, he wouldn't understand. She was feeling her way here and it made her feel young all of a sudden, and small, and she rarely felt young and she never felt small. It made her feel that this time there was a chance she didn't have all the answers. On her own in this bar and with North so far away, she asked herself if she was doing the right thing.

After all, treason was a big word for a small girl.

But she was only small on the outside. And this wasn't treason, she reminded herself. This was a means to an end. Fang slid across a memory key.

'This has everything we have on the Narrow Yett so far. They had a problem going over so he's only just got there.

Apparently, Gaffney originally went crazy months ago. It's all in there.'

Roxie's right hand with its star-spangled nails rested over the key and she slid it off the counter and into her pocket in a well-practised move.

Fang made to climb off the stool. Roxie was dressed in a black trouser suit now, with a silk shirt and small pearl earrings. She wore an expensive perfume, but under it Fang could still pick up the acrid smell of old sweat, as if Roxie had changed but not showered. Roxie smiled, baring the small teeth. 'That wasn't so hard, was it?'

Roxie's doll-like hand reached for the small glass, making it look huge all of a sudden. Fang squeezed her eyes closed against the memory Roxie had shared.

'And are you being a good girl? About the other thing we talked about?' Roxie meant abandoning her enquiries into Benson. 'I'm not a monster. I don't want to hurt you, Fang, I want this to be the start of something special between us.'

It was as if a cold wind passed over her. Roxie was lying. Roxie was a monster and Roxie would be happy to hurt her. Fang had met dangerous women before. Lilith the assassin for one. Lilith was a stone-cold killer. Greedy for money and for the rush that came with earning it. Fang hated her guts. Despite that, Lilith liked her, she knew. Found her entertaining and impressive. Lilith might kill Fang if push came to shove, and she was paid enough. But she'd never enjoy it.

The fact Fang was young and at Roxie's mercy, the fact she had been caught in Lord knows how many crimes, was like blood to a shark when it came to Roxie. Even though she had apparently started to cooperate with her, had just betrayed her country and Hone, Roxie didn't feel any warmer towards her. Indeed, the fact she was capable of treason made Roxie despise her. And now – even more than before – Roxie knew that Fang was in her power.

Fang sighed and it sounded to her own ears as if it had come from the very soles of her Doc Marten's. Behind her Joe 90 glasses, she noticed the lights of the bar reflected in the lenses, and for a second she didn't know which lights were the real light. She glanced around. Out in the foyer she could make out the heavy-set forms of agents she took to be Roxie's backup. As if a teenage hacker would decide to do sudden and fatal violence to a trained CIA agent. She allowed herself a quick fantasy of Roxie's body slumped and dead over the bar, but violence, she reminded herself, was not the answer. The only way out of this mess was to think her way out of it, just like she was always telling North.

'I won't be able to stay here,' Fang said. 'You'll have to get me out.'

'I told you, we have a job all lined up for you with your kind of people. You'll be fine. And after a few years, once you make a contribution to society, you'll be given a new identity and citizenship.'

'And when we go, nobody gets hurt, which means it has to be when North is still away and when Plug isn't around. Plus, I don't want Granny Po seeing it. She's old. You can't do that to her.'

An expression of boredom crossed Roxie's face. 'Let me make your situation clear. You have no rights and no power to negotiate. If I beat you to death with that cocktail shaker right this second, your pathetic government would deny your very existence.

'There are people back home who believe you have real potential to be an asset in our cyber defences.' Roxie smiled, displaying her tiny teeth. 'They think that you're a perfect fit.

Me? I think you're trouble. I don't like children. Never have. Never will. You're too demanding. *Feed me. Put me to bed. Make sure I don't die crossing the road, Mommy.* Want. Want. Want. One less would make the world a better place. And I do get to exercise discretion on the ground, especially if I consider you a risk to US national security. Remember that, cupcake.'

Fang hung her head.

She had to hope she hadn't made a mistake.

30

Murdo

A vicious wind blew in from what felt like three different directions, each one colder than the next. From what he could make out, the island was a strange mix of scrubland, rocks and enormous hills that dropped almost vertically into the sea. Comfort had mentioned a loch and peatbogs, but as yet he hadn't seen either.

Ninety minutes out from the Narrow Yett settlement, however, and a long chain-link fence blocked his path. Barbed wire ravelled out along its top in three rows, and a noticeboard warned: 'Private property. Hazards. Do not enter.' Bearing in mind the island's other occupants were a couple of hundred residents of the Narrow Yett settlement and a few hundred sheep, the Americans seemed to be taking no chances on their messaging. He walked its length. The perimeter fence had regular twenty-metre-tall stanchions with sodium lights attached, but as far as he could tell – no cameras.

Which meant he was free to crawl through the space left as the fencing ran over a dip in the ground. He cursed as the wire caught at his top, using his elbows, knees and feet to move through. Standing, he turned to regard the chain-link, disappearing off into the distance. Was it possible that the contractors had gone to the trouble of fencing off the entire north-west corner of the island? He rather thought they had.

He approached with care, keeping low. The spaceport, which had once been a rocket spotting station, was a much bigger concern than he had been expecting. Sitting in the lee of an ancient wall, he used his binoculars to sweep up and down the site. A handful of low-level concrete buildings, which he thought might have been accommodation and office blocks, with construction underway on more. A bulldozer passed this way and that – its dull growl carrying on the air, construction workers in yellow hats erecting scaffolding against one of the buildings. A silo and a control tower at some distance from the accommodation, with a clear line of sight to what he was presuming was the rocket launch pad. Fang had said there were three of them, so he guessed the other two were beyond the rise in the land, but probably still visible to the control tower. He wondered how close Gaffney had been to the launch pad. Close enough to feel the tremble in his bones? To feel the suck and pull of the air? To smell the burning fuel? Gaffney, too, must have found a way through the fence.

He found the long low building standing apart from the rest and watched it. No one went in or out. He eased his way around and increased the magnification on the binoculars. It had no signage. Was there any chance it was where they kept the rocket fuel? That would have to be kept at some distance from the main accommodation and office blocks on safety grounds. Otherwise, he couldn't account for it.

There was no sign of Kincaid or his men. From what he could make out, the site had its own power station, a fleet of 4x4s emblazoned with the D'Urberville Tech logo, which were parked up close to a massive hangar that he guessed was for their helicopters and rockets. And, of course, its own helipad.

He kept thinking.

There was serious money invested here. No one was cutting corners. Anything with this much activity must have its own medical unit, its own kitchens and dining hall, a shop, entertainment for the workers. Plus, the contractors had to have got their bulldozer on site somehow, along with the 4x4s. And if they were constructing rockets, they were shipping them in at least part built. And indeed the rocket fuel. Not to mention the satellites themselves. Did they use the Narrow Yett's quayside? No commercial operation would have anything to do with the cult. Which meant D'Urberville had to have their own quay and their own way on and off the island via the sea as well as via the choppers. He unfolded the map and the wind almost ripped it from his hands. The most obvious bay that would take a quay was the one he had just left. But to the north-west of the island was a narrow bay that offered some protection at least from the elements. Of course, all things were relative when you were talking about the Atlantic Ocean.

As he watched the comings and goings, his thoughts turned to Fang. He considered his phone call with her. Was it his imagination, or did the kid sound strained? Was she still mad at him for leaving the hospital? Or was something else going on? He could call her back and ask, but he knew what she'd say. That he was a moron-person and needed to get back to the job in hand. Knowing Fang, he had to wait until she was ready to talk.

And much as he resented Kincaid for the fly-by, he couldn't get distracted from the task in hand. He had to shut down the broadcast and neutralize Enoch. And the operation had to be clean and discreet, Hone had said so. He shouldn't push his luck. The last thing North needed was a full-scale brawl with a bunch of Neanderthal security guys. Kincaid could wait.

North eased his way back from his vantage point till he was safe to stand and move away. Once he was through the fencing again, he moved into a slow run. If anyone from the station or from the Narrow Yett were to catch sight of him, he could say he'd needed to stretch out so had gone for a run. Frankly, the wind was so cold, it made sense to keep moving. He kept it up, one click, two, hit the loch at three. A stretch of silver water covered over by wintering geese. As one, they appeared to lift up into the air and take flight, curling and circling overhead, leaving only a handful of swans on the water. For the first time, he felt that perhaps Comfort had a point when she said the island was beautiful.

And Gaffney had been a birdwatcher. He'd probably seen geese on this loch hundreds of times. Probably smiled the way North was smiling. Before he cracked up.

Plus, North had to admit that it felt good to have air in his lungs rather than seawater. He started running again, picking his way through the peat bogs with their dark trenches of cut-away earth, over a stretch of scrubby pastureland, towards the east coast of the island this time.

He stopped again only to get his bearings. As he lifted his head, he gazed around. The sea was ahead of him. The vastness of the sky. Murdo did a good job of reminding a man how small he was in the scheme of things. He'd run further than he realized and the land underfoot was rocky now, only a scraping of grass, not enough to keep a sheep alive. There was no habitation around – no shepherd's hut, no stone-built croft, only a cliff edge and the roar of the Atlantic. And yet he could smell something on the wind. He breathed in more deeply.

Stepping carefully, he approached the edge where the ground fell away, spread flat on the ground, and peered over. The drop enough to make him dizzy. At the bottom of the cliff, vicious rocks, a white shingle beach and breaking waves. But closer, smoke curled up from a narrow ridge of rock. Someone was down there.

31

He thought about calling down but decided against it. Instead, he considered the rock face. If he reversed off the edge of the cliff, and took it slowly, he could climb down to the narrow lip of rock.

Halfway down, North decided he'd made a mistake. Of course, by then it was too late. His boots fought for purchase on the lumps and bumps of rock while his fingers clung to the cracks, the knuckles straining with the effort of carrying his weight. At one point, the toehold of his right boot slipped, and he grabbed at a bulge of limestone in panic, tearing the tips of his nails from his fingertips to hang there for what seemed an eternity but was in retrospect a matter of seconds. A torrent of pebbles came down as he attempted to bring himself closer to the rock, and all the while, the smell of frying fish grew stronger.

'Who's there?' He heard the note of panic before he saw the boy. Rowan gripped what looked like a broken-off oar. He gave every indication he intended to use it to prise North from his already shaky grip on the rock face.

'Calm down, kid.' North let himself drop the last three feet on to the narrow ledge and felt a surge of relief at reaching solid ground. Heights weren't his favourite thing in this world. Rowan blocked his entrance into the cave. A gust of wind hit the back of North's neck. Last time he'd been in this position this high up and this precarious, he'd ended up dropping into the Thames. This time if the kid pushed him off the ledge, the jagged rocks below would make a comeback impossible.

He held up the palms of his hands as if to show the boy he came in peace and took a step away from the edge, and then another. He kept his movements slow. The kid seemed twitchy as all hell. 'The best thing for both of us is you go back to your pan before that fish burns and I have a few moments to think about spending the rest of my life in this cave, because I have no idea how I'm getting back up there.'

Rowan stayed where he was as if reluctant to retreat, because retreat would mean allowing North into his territory.

North smiled. 'Kid, seriously, give me a break. I don't have it in me to climb back up right this second.' As he said it, he realized it was true. His legs trembled and it hurt to breathe. Not from the run – his body was remembering the hurts that meant he should really be resting up in a hospital bed. He set aside the pain. 'You have to be part goat.'

The boy glowered but stepped back. With some relief, North bent his head and followed him. In the lee of the entrance to the cave, a fire burned in a small scoop of rock, more rocks piled around it to keep the wind from blowing it out. Over the flames, a crisping fish was frying in a heavy skillet with what looked like a few oats scattered over and around it. The boy settled down on his hunkers and shook the pan as North took up a place opposite him. Keeping his look casual, North glanced behind them. It was dark, but even so he could see the light and dark layers of grey and sparkling rock and that the cave appeared to be lined with what he took to be the arching ribcage of a whale. It was one of the strangest things he had ever seen. The boy was the size of a twig. He could not have been the one to install the bones. Which meant they had to have been here decades, if not longer.

Rowan's eyes were on him when he turned back, a look that dared him to pass comment and find his surroundings wanting. North nodded. 'I like what you've done with the place, Jonah.'

The boy's shoulders relaxed, his expression still wary but lit now with a flicker of what might have been a smile. 'You know the story of Jonah and the whale?'

'The fact he was swallowed whole?' North said. 'And spent three days and three nights in its belly, before it threw him up? I always wondered if afterwards, the whale went around telling all the other whales about the one that got away.'

Rowan gave a sharp laugh. It didn't sound like he'd had much practice.

The cave went back further than North had expected but he guessed the wind over the sea kept the air in the cave fresh and the boy safe from CO poisoning. What the boy did for water, he didn't know. Presumably there was a stream towards the back? The island certainly had plenty of water – he'd seen the well heads marked on the map.

Was this a den to keep him out of Drummond's way during the day? Because, with peat banked by the fire and drying seaweed draped over the walls, it looked way too elaborate for that.

As a kid himself, North had spent enough nights on the streets to know the value of shelter and Rowan had done his best to make it comfortable. From somewhere, he had scavenged a threadbare rug, which he used as a bed. Three blankets were folded neatly and laid on the rug, along with what looked like several old coats and a shredding tarp; a plastic bucket held what North took to be fresh water and a few meagre supplies were stored on an old tea crate, along with a battered leather-covered book and a camping lantern.

'You live here all the time?' he asked.

Rowan shook his head. 'In the winter if the winds are bad, I go up to the burial mound at the north of the island. Have you seen it?' He didn't wait for North's answer. 'It's five thousand years old. But that's nothing. Murdo is built on rock that is fifty million years old. Isn't that amazing?'

'That's older than me,' North said, and the boy smiled again. 'Don't let me keep you.' North gestured towards the crisping fish.

'Do you want some? There's enough.'

North was hungry but from his companion's hollow cheeks, Rowan needed it more. He shook his head. 'You go ahead, kid,' he said. North sat cross-legged on the stone floor, the boy across from him. A draught from somewhere behind them pushed the smoke out of the cave. The boy extracted a penknife and used it to break apart the fish in the pan, letting the juices seep into the oats, before lifting the charred skin and barely cooked flesh to his mouth with grubby fingers. North sat peaceably. Rowan needed to get used to his presence, realize he wasn't a threat.

'I didn't need you interfering.' The wary brown eyes switched from the fish to North.

'I know you had it handled. But he was a big guy.'

'I wasn't kidnapping her.' The boy took another mouthful of the fish.

'I'm sure you weren't.'

'I was saving her.'

'Good for you.'

It took barely a couple of minutes before the fish was eaten. And with a sigh, as if the fish had been good but he still wanted more, the boy put the heavy pan down and wiped the knife clean with a handful of sand from the floor of the cave. Then he let the sand trickle between his fingers on to the black metal, using a piece of sacking to clean the surface. North liked that. It showed discipline.

'Tara's my sister... my half-sister,' the boy corrected himself carefully. 'That means it's all right for me to take her someplace safe.'

North nodded, as if the boy was doing no more than stating the obvious. It was like talking to a wild animal, the russet hair, the dark eyes, the nervous disposition, and the boy himself seemed unfamiliar with talking, although now he had started the words began to pour from him. 'You don't know what it's like here. Drummond and his men are thugs and the rest of them are lunatics or go along for an easy life.'

'Tell me about Drummond.'

'What you did. Knock him flat like that. He won't let it go.' The boy let out a heavy sigh. 'He's going to kill you.'

'He can try.'

Rowan hung his head, shaking it from side to side as if he didn't want to go there. 'They want me out of here. And that's all right with me...' North could see the hurt on the boy's face. How long had he been living outside the community? Living in this cave? From the boy's skinniness and fragility – too long, he suspected. 'I don't need any of them. But I'm not leaving my sister with them. I was hiding out with her, trying to get to the boat without them seeing, but they found us.'

North said nothing – he needed more.

'Drummond is her father?'

Reluctantly, the boy nodded. 'He doesn't care about her really. My mum – Lisette – she'll have made him come after me. She can be a real... She's his... She's into Tara, she just doesn't love…' The boy was too loyal to the idea of his mother, the memory of who she had been, or who he had wanted her to be, to say the words, which North could have said for him. Rowan's mother wasn't up to much. She'd washed up here with her boy years before, maybe addicted to drugs or drink, maybe addicted to running away from the real world, which meant facing up to the need to put a roof over their heads, and paying the bills and caring for her child. Drummond was just the latest in a line of unsuitable men who offered her a warm bed and someone stronger than she was to take charge. Surrendering herself had suited her fine. But it seemed Rowan wasn't minded to do the same.

'Drummond's a headcase and he's in charge here…'

North interrupted before he could help himself, forgetting that he was a stranger and should know nothing. 'I thought this place was run by Enoch.'

Rowan nodded. 'Sure, Enoch does the talking and the praying, but Drummond organizes all the computer stuff and runs the money side of things. Why are you here? Did you see Enoch on YouTube?'

The corner of North's mouth twitched upwards at the suggestion he could be that naive. He didn't want to lie to the boy, but he couldn't tell him everything. 'I got in a fight back on Harris. I needed someplace to lie low. Tell me something – why aren't you like the rest of them, Rowan?'

The boy swelled at the compliment. He shrugged his bony shoulders up and down as if to say he didn't know, but his eyes went to the leather-bound book on the crate.

'What's the book?'

'It's a journal. My mum and me, we were one of the earliest families here. The church has taken over all the cottages now, but for a long time there weren't many of us. The people who used to live here left nearly a hundred years ago.' The huge brown eyes were distant now, as if imagining the boats pulling away from the island loaded with islanders and all they owned. 'They left behind a Bible and a handful of oats in each cottage. I guess it was like a blessing. But the schoolteacher was sick. I don't know what with but it's all in his journal, he was coughing up blood. He was supposed to be on the last boat out of here. He'd packed up his clothes and he'd crated up his library including that journal, wrapped them in oilskin, and then stored the crates in the outhouse. There were too many for the house, he said, and it made him sad to see them. But before the boat came for him, he took to his bed and by the time they came, he'd died. They buried him – there's a marker in the churchyard, but I think they must have just shut the door on his cottage, and they can't ever have checked the outhouse.'

Rowan reached out to his bed and lifted the corner of the rug. It was laid on stacks of books, forming the strangest divan base North had ever seen. He leaned towards them. He couldn't make out the titles of most of them, but he could see Tolstoy and Chekhov, Sir Walter Scott and Shakespeare, the Brontës, poetry, mathematics, philosophy and theology. Some of them were in Latin.

'I went into the outhouse one day, just nosing around, and opened up the crates. The door was locked and the roof had half caved in. It looked a wreck from the outside. I let myself down into it through the hole in the roof. It wasn't stealing...' His voice was defensive. 'We're not allowed books, but I love reading...' Of all the things the boy had said so far, this was the thing that surprised North the most. 'I had to find a safe place for them or Enoch would have burned them. At first, I kept them here, but when Drummond got too much, I just moved in.'

'You're serious – they'd have burned them?'

'Yeah. The kids are taught to read and write and code, of course, but they're only allowed to read what Enoch writes for them and the texts on screens that he says are okay. He says it's all about keeping us free of contamination so that we're viable in another simulation.'

North looked back at the books. The kid had taught himself to think. And thinking for yourself in a cult was a dangerous habit, even if he was reading books that were all written before 1929.

'I think this schoolteacher would have thought they'd found a good home,' North said. It hurt him to see how easily the boy was pleased by the words. 'What about food and water?'

The boy gestured to the back of the cave. 'There's running water. I collect eggs. I fish – I made my own net – and Mia brings me food. And sometimes...' He hesitated. North could see a flush rise in the pale cheeks.

'...you steal,' North said. There was no judgement in his voice, because he wasn't judging the boy.

'I check the trash and I can make soup. When I run low, I take scoops of lentils and beans and rice from the kitchens and the cleits.'

North looked blank.

'They're the stone huts with turf roofs. People who used to live here used them to store food and peat. *They* do too.' North noticed the boy said 'they' not 'we'. 'If I can, I take oats and flour and eggs. They have chickens down there.' He said 'chickens' the way some kids would have said 'unicorns'. 'Not so much that they'd know anything is missing. It's enough.'

'Do you ever go walkabout in the tracking station?'

The boy nodded. 'But they burn their waste, so I only do it

when I have to. A few of the guys give me chocolate bars and cigarettes when they see me.'

'Really?' North raised an eyebrow. So the Americans would leave him to drown but feel sorry for a kid out here on his own. 'Ever see a guy with a burnt face?'

Rowan nodded. 'Yeah. Not him – he's mean. His name's Kincaid. He caught me inside once and said if he saw me again, he'd "whip my arse raw".' That sounded more like it, North thought.

'Where were you when he caught you?'

Rowan shrugged. 'Just poking around.' North wondered if Kincaid had overreacted because the kid was near the mystery building Fang had mentioned.

'There's a long low building that stands apart from the others. Do you know which one I mean?'

Rowan nodded.

'Do you know what goes on in there?'

Rowan looked thoughtful. 'I don't know. It's got no windows.'

'How about the rocket launch? Did you see that?'

The boy grinned. 'It was a-mazing!' he said. And suddenly Rowan didn't look hungry or desperate. 'The earth shook and fire came out from it, and it took off – whoosh!' He lifted his arm into the sky, a smile from ear to ear. It was infectious. The kid was a natural enthusiast. 'I wish I had a telescope.'

'And how about when you're looking for food back on the settlement. Ever been caught by anyone there?'

Rowan's face turned serious again. 'I know my way around the kitchen. They use food as a reward and a punishment. If Enoch's pleased with you, you get proper meals. If you break curfew or they want you to do more work, you go hungry and they call it a "reality mortification", but they say it like it's a good thing. Drummond says they're all idiots.'

'Isn't he part of it?'

'Not always. There are ceremonies where you're half drowned in the sea and Enoch calls it a "reality baptism". Drummond doesn't go to things like that. And there are ceremonies with snakes even. Real live snakes. One of them bit the guy who used to look after them and he up and died. Most everyone has a tattoo of a gate with a snake like a "C" around it. The tattoo isn't supposed to be compulsory, but it totally is. Drummond and his men don't though, and they have all kinds of tattoos. If you do something and Enoch doesn't like it, or one of the creeps around him don't like it, no one is allowed to talk to you for weeks or months at a time. You're supposed to enter "Reflection" – that just means they want you to feel bad. Or they make you do really bad chores, like just one person has to clean up after everyone eats, which takes hours. They try and humiliate you till you agree to what they want. Or they talk and talk and talk at you till you want to scream. And if someone wants to—' He paused, a flush starting up on his cheeks.

North kept his own face neutral.

'—wants to sleep with you, you have to have a really good reason why you won't do it. Drummond and his cronies like that side of things here.' North thought about that. About Comfort and Joy having to sleep with Drummond and his thugs, because the men wanted them. Not given the right to exercise choice. It would account for Comfort's fear of Drummond. The dislike he'd sensed.

'But there are visitors over from the mainland. Surely they wouldn't stay – they'd run a mile?'

Rowan shrugged. 'Visitors only come in summer. They stay

for a while and then they go. If the visitors notice someone isn't eating, they're told it's a detox.'

That made sense.

'How long have you been out here on your own?'

Rowan thought about it. 'About three years. I lose track.'

'Was it really bad a year ago?'

Rowan looked surprised that North knew so much. 'Yep. Drummond and his men all went across to Glasgow. There was some kind of trouble, I think. I don't know exactly what happened, but I went down to try and get some food, and they were all off their heads partying. I snuck in and snuck out, but for weeks after there was screaming and it felt off. When Drummond got back, he and Enoch had a huge row about people hurting themselves and "liability". Everyone was talking about it.'

Rowan was smart. He'd worked out a lot of stuff by himself – like what was right and what was wrong. It must have made for a lonely growing up.

'And now Mia's here? You said she brings you food.'

The boy's face lit up again – he was a superfan, it was written all over him. He puffed up with self-importance. 'She brings me cold meat and eggs and fruit and bread and she doesn't care who sees her. Sometimes if she can't get extra, she brings out what's hers to eat and we share it.'

North wasn't the only teenager to go hungry. Mia had a background as bad as his own. The neglect and abuse of Rowan would enrage her, or would have enraged the old Mia. Why was she still there then? Why was she working with Enoch to churn out propaganda for the cult? Or was she planning to turn them over? If it was as bad here as Rowan said, she would know all of it. Mia would never be anything other than good at her job.

The thought of Mia and the settlement reminded him that by now Comfort would be looking for him. Mac had been keen to get away. He was safe from being thrown off the island and into the sea but if he stayed out too long, Comfort would wonder where he'd got to. He glanced at his watch. 'Come back with me and I'll make sure Drummond doesn't get anywhere near you. I promise,' he said. If he had the chance again, he'd knock the guy's head clean off of his neck.

Rowan shook his head. 'You don't understand. Drummond doesn't believe anything Enoch says. But even so, he does what he wants round here. He knows Enoch from when they were at university together. He and his men run the place. They're back and forth to Glasgow running errands for him, doing business, but he always has a bunch of them with him on the island. The younger ones work out of the old church, where they have all the computers and where Enoch broadcasts from. I've heard them talking. They've done bad things.'

'Bad things?'

The boy's eyes were huge. His voice fell away to a whisper. 'You should stay with me. Until the boat comes back, then you could help me get Tara away. The captain would listen to you. Drummond wants out. He's never let his men mix with Enoch's followers apart from the women they sleep with. He says they're all "flakes". I heard him tell my mother that Enoch's getting "crazier by the minute". But if he leaves and takes Tara with him back to Glasgow, I'll have lost her forever.'

The boy's worry was genuine, as was his realization that North was his best chance of escape with Tara. And he had that right. Because North wasn't leaving this kid on an island of lunatics.

It was easy to think he knew the boy, not least because he'd been this boy. Not in a cave on a rock in the middle of the ocean. But in any number of dumps in the middle of any number of housing estates across London. North's childhood had been grim. Growing up too fast because, in the grip of addiction, his mother stayed a child. Coping with the fallout of the losers, the grifters and the violent predators that fed off her. He'd dragged himself up. Got himself up and out and to school. Gone hungry more often than he cared to think about. Had scavenged in dumpsters behind shops for discarded food. And when he had to, had shoplifted food and soap as much to feed his mother as himself. He'd been beaten by violent men, slapped and punched for no better reason than he was there. He'd spent nights sobbing and curled up in fear till he learned not to cry. He'd been scorned and excluded by teachers who should have known better. He'd been a problem social workers couldn't solve, passed around the care system like unwanted garbage. North knew exactly how neglect corroded a child's sense of self-worth. How happiness was a luxury an abused child could only dream of. That hunger and violence made for dark companions. He looked into Rowan's eyes and saw the hurt there. Took in the thin face and the scrawny arms and felt the hunger in his own belly start up. Glimpsed the bruises Drummond had inflicted only that morning and felt the tender ache of them in his own skin.

The wonder was that Rowan was still whole and righteous. Strong enough to leave. Smart enough to find shelter and feed himself. Brave enough to try and carry his sister to freedom. What would happen if North left Rowan here in his cave, on this island, in this ocean?

The best-case scenario, Rowan survived. He didn't thrive. Survived long enough till Drummond got him off the island and Rowan took to the streets. A young boy on his own. More likely, Drummond would hurt him badly or even kill him. North could judge a man's capacity for violence with a glance. Drummond was a beast. Clever and savage. One night, Rowan would be caught in the kitchen and dragged in front of Drummond. North couldn't see Rowan pleading for mercy. He'd dare Drummond to do his worst, and on an island far from laws and civilization, Drummond would do just that. He would beat the boy senseless and throw what was left into the sea to drown. For a year or two, Rowan would be a memory. And then he would be nothing at all.

North had a job to do – stop the broadcast and stop Enoch. But that wasn't going to get in the way of saving Rowan and his sister. Getting them out of here and to someplace safe. A place where a child could be warm and fed and loved. As a boy, North had been left lonely and had no choice but to turn to violence. It had been the only way to survive. No one had been there to save him. No one had wanted to save him. No one had thought he was worth saving. That wasn't happening to Rowan. North was here and North was saving the kid.

His blood started pumping faster. The headache he'd been nursing since slamming into Tower Bridge faded into nothing. He felt sharp and he felt dangerous.

He had plenty of questions about this cult. Like how a man came to be castrated. How a crime lord managed to launder money through it. And how a boy called Rowan was abandoned to life in a cave full of whale bones. And the old church seemed like a good place to start looking for answers.

32

Rowan guided North back up the rock face, which transformed the journey upwards from potentially life threatening to plain and simple terrifying. He hadn't liked leaving the kid there. Drummond must know where he was. Even so, the kid was adamant he was safer on his own than in the village.

North considered the issue as he made his way back to the settlement. Something was off about this whole thing. Was Drummond exploiting an old friend? Capitalizing on the opportunity the Narrow Yett provided to make millions? Did Enoch trust his friend of decades? Or didn't he care about his friend's crimes, providing he was left to abuse and control the members of his 'church'? After all, Drummond might beat a man to death with his bare hands. But money drove men like him. And North couldn't see any money for the gangster in the castration of lost souls like Gaffney. He couldn't see money for Drummond in keeping members of the cult hungry and compliant. He couldn't see money in cutting them off from their families or in the complaints of those who made it off the island.

Rowan had said Drummond ran the cyber operation of the cult, and oversaw the finances of the place. Other than that, he and his men kept their distance from 'flakes'. And according to Rowan, Drummond believed Enoch was getting 'crazier by the minute' and they had rowed badly around the time Gaffney mutilated himself. Drummond was walking a fine line. It was in his interests to keep the Narrow Yett popular enough to bring in donations for his money-laundering scheme to work, but far enough away from controversy that police and lawyers didn't start asking questions. He'd even refused cooperation with Mia's documentary. He wanted a low profile. And there was nothing low profile about Gaffney's atrocity.

There was a light flickering in the cottage. Could a heart really skip a beat like that? Apparently, it could. Mia must have come looking for him. Wanting an explanation for his presence on the island. An apology for his silence of fifteen years. Wanting to see him. He was grinning as he opened the door. But his smile went as quickly as it had come.

Comfort was in his bed. And she had the look of a woman who was beginning to wonder whether she'd made the right decision when she stripped naked. 'About time,' she said, as he closed the door. But she sounded nervous. He kept his eyes soft as his gaze took in the room, sweeping the rucksack, the clothes drying by the fire, and the shadowed beams. The T-shirt had gone from the rucksack and was instead hanging with his other clothes. Had she done him a favour or had she searched the rucksack more thoroughly this time?

'I usually expect dinner and a movie first,' he said, keeping his voice even as he came further into the room. The disappointment he felt was coruscating. He was an idiot for thinking Mia would be waiting for him. Still less in his bed. He was the one at fault, and she'd be expecting him to seek her out.

He stripped off his jacket and hung it by the fire. Drawing out a chair and sitting down at some distance from the beautiful girl curled up in his bed. Comfort turned on to her side. The

SLEEP WHEN YOU'RE DEAD

firelight dancing across the skin of her naked shoulder – the blue hair spread out on the pillow transforming her into something unworldly.

'Where've you been?' She seemed half asleep; even so, there was an edge to the question.

'I didn't want that guy Drummond shoving me back on Mac's boat,' he said. 'Seemed safer to clear out of here for a couple of hours.'

She looked as if she was considering his answer. 'He said you're to keep out of his way. He made a big deal of the fact he was doing me a personal favour. He's a brute but he can be charming when he wants to be. If it suits him anyway. He says I'm wasted here and he could get me a job managing one of his nightclubs. That during the day, I should go to college and study fashion or media or something.'

'And are you considering it?' The more he heard about Drummond, the harder he thought he should have hit him. He had no doubt now that Drummond had demanded Comfort's presence in his own bed. And that she'd felt obliged to go there. He didn't believe a word about managing one of the gangster's clubs. Drummond would put Comfort on the streets and there was no way North was letting that happen.

North moved to the edge of the bed and stared down at the guileless face. 'Men always think because I'm young and pretty that I'm dumb,' she said, twining her fingers in with his. 'I'm not dumb. You get that, right?' He nodded. She didn't smell of patchouli any more, but of soap and the smoke from the fire. Flames flickered in the blue eyes.

The possibility of what could happen between them was real. And he was tempted. Comfort was a vision, and sweet with it. Who wouldn't want to sleep with her? If he hadn't already come across Mia on the island, maybe he'd have done just that. But Comfort had been exploited in this place. She'd been trained to believe men had the right to demand sex of her and she had no right to reject them. North wasn't going to take advantage of her innocence. He liked her too much to do that.

As he thought it, Comfort sat up in bed and pressed her lips against his, which involved pressing the rest of herself against him too. Instinctively, his hands went to her, feeling her narrow ribcage, one sliding to the small of her back, and she let out a little gasp at the touch.

He wanted her. He couldn't pretend he didn't. But what might happen here – wouldn't be fair on Comfort.

He held her away from him and in the firelight, the huge eyes looked apprehensive. 'Don't you want me?' Her voice was tremulous, and she looked like she might cry. 'You're the first man in a long time I've wanted to be with. Please.'

But he wouldn't hurt her. And after the flirting, his refusal could look bizarre. 'We could have died out there today. That can mess with your head.' Much like a bullet, he thought. 'And to be honest, without wanting to come off like a complete wuss, I'm exhausted. How about we get something to eat? We get some sleep tonight and we revisit this tomorrow? Or the next day?' He opened his hands wide, palms upwards. 'I'm not going anywhere.'

Something flared in Comfort's eyes as she lay back against the pillow. 'No, I guess not.' She gave him the kind of smile that made him want to take back every word and just kiss her. He held tight to the idea of Mia.

33

London

The temperature in the Downing Street office was arctic and yet the windows were open, or perhaps because the windows were open the temperature was arctic. Either way, Eugenie Webber showed no sign she felt the cold. She was a big-boned woman with thinning hair retreating at the temples and a pudding-like face devoid of make-up, a fact the tabloids mentioned in every feature they wrote about her, as if to explain to readers that nothing could be done about the photograph they printed alongside the articles. If she had had any say in the matter, she would have kept her name out of the headlines, but as the prime minister's chief of staff and most powerful agent of change, it was hard to manage.

She and the PM had been at university together. There had been rumours of an affair at the time, but no one gave it much credence, although as it happened it was true. Three months of sado-masochistic sex with her as the dominatrix and her lover as her slave. Three months when she could reduce her undergrad lover to a gibbering wreck by a snap of her vinyl-gloved fingers. And three months in which they made a pact to make him PM and for her to remain at his side throughout. Neither regretted the alliance and she knew the PM still thought about the sex. She knew it because occasionally she would snap her fingers on some pretext or other, and he would flash her a dopey grin.

No one else, however, wanted to think about Eugenie Webber having sex, not even Eugenie herself. She'd tried it with all the usual variations – not just the S&M – and with a variety of partners throughout her twenties and found it underwhelming, although she'd pretended otherwise. There'd even been a brief marriage to an elderly mathematician. But even when young, she believed there was more satisfaction to be had elsewhere.

The man across from her had already turned down the offer of coffee. He'd kept on his long waxed riding coat, which was an affectation, she thought. Then again, he was an intelligence agent, so perhaps he knew she kept the place cold. Or perhaps he was keen to keep the meeting as brief as possible. As usual, she'd attempted to steer him towards the orange mohair sofa, which had the consistency of a marshmallow, but he had swerved at the last minute and taken the chair across from her desk. The sofa served two purposes. Those seated in it were forced to look up at her, and it gave a homely feel to her office. It was in fact more homely than her actual home, where she preferred a stark minimalism. But here the chintz cushions and the family photographs, the fresh flowers and the cartoons on the wall, converged to lend her some humanity and warmth. She even ran to a scented candle on occasion. In reality, she had long since accepted that she was both less than human and more than human. 'Less than human' in her lack of emotions and relationships outside of her family, and 'more than human' in her forensic clear-sightedness and ability to think differently to the hoi polloi. Commonplace emotions such as romantic love or hate, jealousy or envy, sadness or joy, were, she believed, at best a time-wasting hindrance and at worst downright destructive.

But ever since the suicide of Jonathan Gaffney on Tower Bridge, human emotions had come roaring back into Webber's substantial bosom. Right this second, the emotion she was feeling as she sat across from Edmund Hone, head of the MI5 section colloquially known as the Friends of Cyclops, was a quirky combination of bleak rage and absolute blood-pumping desire. It had been a long time since she'd felt either and so unfamiliar were the sensations that they had brought her out in a light sweat, which had required her to mop her face and palms dry twice already.

Hone affected not to notice, but she knew he had, and she allowed herself to resent him for it. The effect of one emotion crowding in on the others had a similarly intoxicating effect, she thought, to the single dirty martini she allowed herself every evening at 6 p.m.

'I've read the briefing. What's not in it is the fact this Tower Bridge atrocity is something of a clusterfuck, wouldn't you say.' It wasn't a question. She smiled with heartfelt insincerity at Hone as she leaned towards him. He didn't return the smile and he didn't ask her permission, merely extracted a silver cigarette case from his pocket, flicked it open, selected one, and with his thumb and middle finger placed it with some care between his lips. She readied herself to object. Instead, she realized she also wanted a smoke.

He removed the unlit cigarette but kept it in his hand. His stillness had an animalistic quality, she thought, as if he had spent years of his life waiting in the shadows before moving in for the kill. They had that much in common then.

'And now the perpetrator's wife…'

'Ex-wife.'

Eugenie glared at the one-eyed man but he appeared oblivious. Merely contemplated his cigarette as if willing it to light itself. She carried on. 'And now his "ex-wife" is claiming Gaffney was disturbed and that our surveillance exacerbated his condition, placing responsibility for his subsequent violence on us rather than on him.'

Hone's face remained impassive and as ever, she tried not to stare at the jagged scar that ran the length of his right eye socket. She had once demanded a security briefing on Hone just to satisfy her curiosity about the eye. Ninety-nine per cent of it came back redacted.

'Prior to our pursuit, Gaffney had gaffer-taped two carving knives to his hands. Knives he subsequently used to cut his own throat and all but chop off his own head. I'm not sure anyone could put that on us.' His voice was calm and reasoned. But this wasn't a man used to being held to account. And the faintest taint of insolence in the tone made her want to slap him. Men like this thought they had the right to make all the decisions, but not any more. She could fire him if she wanted. She could make life so difficult for him, he'd be forced to quit. She could— She checked herself. It was as if every emotion she had failed to feel over the last ten years wanted a spin out before she clamped them back down. Control, she reminded herself. She had to stay in control of herself to stay in control of men like this one.

'And naturally the media is already questioning the nature and quality of surveillance, bearing in mind what went down,' she said, a note of fake regret in her voice.

Hone shrugged. 'Doubtless, we'll tell them "lessons have been learned".' A smack of 'who cares' in his response.

'I'm sure they'll be pleased to hear it.' It was like a fencing match, she thought. The cut and parry of it.

Somehow, Hone had managed to light his cigarette without her noticing. She bridled at his effrontery but decided against making him put it out. Instead, she opened a drawer and extracted a glass ashtray then a cheroot from a silver box. She lit the cheroot with care, taking some pleasure in blowing the acrid smoke in Hone's direction. She had the power here, not some jumped-up spymaster, she reminded herself. But she had to play this right. Hone was an operator and this was what she had been waiting for. The reason for all of it. For abandoning academia and going into politics. And for putting up with vainglorious dolts like the prime minister.

'These Narrow Yett people. I don't like any of it. What are your thoughts?' In her experience, men always loved to be asked what they thought. She rarely listened. But this time she did. Because this time she had 'skin in the game', as her father used to say as he read the financial pages over breakfast. An expression that always annoyed her mother. And over the pink pages, her father would wink at Eugenie and her sister, and they'd giggle. Now, Eugenie sat forward in her chair, her elbows on the desk. She placed the cheroot in the ashtray and laced her fingers together. Attentive. Ready to be advised – providing Hone advised what she wanted him to advise.

'It's a cult,' Hone said. His eyes went to the cheroot as if he wanted to smoke that too. 'The simulation theory they espouse allows for some intellectual posturing. For some, the spiritual side is out there but genuine enough, as genuine as any of these things are, anyway. It has an environmental bent and there's a lot of talk of community and social responsibility. But for most, it offers a validation for their desire to operate outside social norms and a way of perpetuating various outlandish conspiracy theories.'

'Conspiracy theories? You mean Gaffney didn't believe in this reality so he didn't believe he had to follow our rules? I'm guessing that's how all the members of this cult feel.'

'We don't know that yet, but it's possible. As you know, we understand Enoch is planning to broadcast a call to arms. Obviously, we need to stop that happening.'

Eugenie sniffed. 'Hone, trust in government is at an all-time low, not helped by those conspirators who tried to bring down democracy last year, or that spate of assassinations we had soon after. For God's sake, look at that atrocity at the British Museum. Bodies everywhere. And you and I know, we've managed to keep the worst of it out of the headlines. Even so, it's been one thing after another. I do not want more bloodshed on our streets. I'm not risking any QAnon situation over here. Shamans running amok in Parliament. We are in desperate need of stability.'

It was an explanation of sorts for the conviction.

'Murdo is a priority. I want the New Army sent in. Special Forces. If Enoch Fraser gets killed during his arrest, that's all to the good. I want the Narrow Yett as an organization taken apart, before there's more Gaffneys running amok with kitchen knives. One man can't do all that, even one of yours.'

'You haven't met him.' Hone offered the information as if making a concession.

'I've met his type. Dear God, Hone. Think about it. There are three hundred people up there. How are you expecting one man to take on an entire cult? It's impossible. In my opinion, the Narrow Yett poses an immediate threat to national security. Islamic terrorists, right-wing extremists, incels, we have enough of these people out there already.'

'I'm going to tell you what I told him. These people resent authority. If we send in the New Army before we take down Enoch, Murdo could go up in flames. There's women and children up there. It would be a PR disaster for you.'

She knew he was trying to work her, mentioning public

opinion. Hone didn't give a damn about public opinion. She turned the full force of her displeasure on him. 'You take the word "intelligence" away and you're just left with the word service, aren't you? As in – who do I complain to about the service?'

Hone's one eye watched her. She couldn't make out the colour of his iris. If she'd had to say, she would have said it was black.

'Would your opinion be influenced at all by the fact the Americans would rather like the island to themselves for their little spaceport?'

Eugenie's face was a study in dispassion. These men always presumed they were so much cleverer than she was. She had her reasons for what she was asking for. And they had nothing to do with the Americans.

'They've spent a lot of money up there,' Hone carried on. 'They must have friends in high places to have been able to push everything through so fast. Don't they get on with their neighbours? Do they shake their tambourines too loudly on a Sunday?'

She spoke slowly. As if she was holding back her temper. 'You know better than most, Hone, that it's as well to stay on the good side of the Americans.'

There was a pause as they regarded each other. Eugenie's voice dripped with honey as she leaned forward, her substantial bosom resting on the desk. 'There are obvious advantages to having an operation such as your own. Scrutiny can be so awkward when it comes to operating within the rule of law. And, of course, the "Friends of Cyclops" have an outstanding record, which lends you a credibility and authority you may not otherwise have. But you operate at vast expense, and you are only useful for as long as you are useful. Not one moment longer. Are we clear?'

Sighing, Hone raised an eyebrow and stood, as if to say he'd heard it all before and he presumed the interview was over. Had he taken offence? She didn't care. 'And one more thing, Hone. The prime minister's wife...'

'... has done an online Narrow Yett consciousness manifesting course at some expense and Enoch attended a function at Downing Street last year. I understand there are photographs in circulation.'

Eugenie appeared to wince. In her fads and friendships, the prime minister's wife was a constant source of embarrassment to the PM. 'The photographs are no longer an issue.' Because she had bought them from the newspaper editor in possession of them, whose wife's dearest wish was a knighthood for her husband. Both the photographs and the camera's original SD card were now in Eugenie's safe deposit box. 'But I thought you should know. All right, Hone, we'll do it your way. But if I don't get Enoch's head in a box tied up with ribbon, your entire operation is over. The Friends of Cyclops will be dispersed to the four corners of the civil service, never to be spoken of again.'

She ground the cheroot out in her ashtray, enjoying the stink of it, and by the time she lifted her eyes, Hone had slipped away – the door to her office opening and closing silently – and only the smoke from his cigarette proved he had ever been there.

34

Hone contemplated the traffic as it rumbled up and down Whitehall – the black cabs, the red buses. His cyber experts had spent some time giving him a crash course in simulation theory. He couldn't see the attraction. No simulation could construct the idiocies, the bleakness and the death he had been witness to. It was all too complex and too unnecessary. No, Hone was a firm believer in reality, good and bad.

Rush hour was beginning to build. He stood a little way down from Downing Street. He should get back to the office, but he needed a moment to think about the scene that had just transpired. Out of the corner of his solitary eye, he noticed one of the armed police officers watching him, until a colleague crossed over to him to explain that he wasn't a threat. Not the kind of threat they were looking for anyhow, and, satisfied, the officer moved off to glare at a group of tourists.

Eugenie Webber was not to be underestimated. There were those who considered her the most powerful person in the country – more powerful even than the PM, rumoured to be in thrall to her intellectually and emotionally. But demanding the New Army went in seemed at odds with her reputation for arctic professionalism and subtlety.

A poster caught his eye, taped to a lamp post. The black-and-white image of a young woman, an MP assassinated close to the Commons last year. A victim of right-wing extremists. The remains of a small burnt-out votive candle at the base of the lamp post. There had certainly been enough mayhem in the country in recent months. The country didn't need more extremists and bloodshed. He'd stood his ground, but was Eugenie right to demand boots on the ground up there? Perhaps she was merely trying to get ahead of the problem?

He kept walking down Whitehall and towards Parliament Square.

Maybe North might not have been the best choice after all? He was up there on his own on an island in the middle of an ocean with three hundred people devoted to a man he might well have to kill. Hone couldn't see that going down well. And he would still have to get off the island afterwards. Perhaps it wasn't the army so much as the navy they should be sending in?

And leaving aside the forces ranged against him, North had a bad habit of being in the centre of things when it all got messy. Then again, that was part of the attraction. One of the reasons Hone had recruited him. For a young man, North had a cool head, a ruthless streak as wide as he was, and a commitment to extreme and fatal violence when he judged it necessary. It was a winning combination as far as Hone was concerned. Plus, he was about as disposable as they came.

Hone understood the impact a childhood like North's could have on someone. As a teenage boy, when faced with the choice between being a victim and a killer, he'd chosen to become a killer. Certainly, it had been a risk to recruit him after the years North spent murdering those enemies of the state that the Board identified. But Hone had tracked his work and taken no issue with the targets or with North's effectiveness. He'd presumed he was dealing with a psychopath lacking both conscience and a moral code – a puppet at the beck and call of ruthless men. But he knew now that North was no psychopath. He was intelligent and thoughtful and loyal. A romantic when it came to women and someone who inspired devotion among his band of misfits.

Cross him, however, and there was no mercy. The savagery with which he had despatched the Board after he realized they'd killed the woman he loved had given even Hone pause for thought. But he'd come across very few people with North's skill set and – he'd give him this much – sense of duty.

The problem wasn't the bullet in North's head. The bullet was a convenience for Hone, and he'd been honest with North about that. The problem wasn't even that North attracted an eclectic – albeit talented – group of helpmeets who owed Hone nothing. No, the only problem was North's insistence on thinking for himself. Agents were like dogs, you had to get them when they were puppies. And young though North was, Hone knew he was never going to come to heel.

This operation needed someone who could improvise. Someone cunning enough to make it onto the island and remain credible, intelligent enough to figure out how to stop the broadcast in time, and ruthless enough to kill Enoch without compunction if that's the way it went. Of course, soldiers could go in flash-bang, kill Enoch and haul away his followers. There'd be casualties, a lot of them. But it was doable. Soldiers, however, could do nothing to control the Narrow Yett's online followers. They were the ones who would have to turn their back on the cult and accept the reality of reality. For that alone, he needed North on the ground to make a call on what could be done.

As for everything else he had going on, this particular stratagem was a complicated one. Beyond complicated. It was insanely ambitious. He'd admit that to himself, at least. What he really had going on was a thing of genius. So audacious, it was a play that would go down in the history of the security service under an index listing that read 'Madness'. And he didn't need politicians and their flunkies getting in the middle of it.

Eugenie's threat was overt. Why? She had her own agenda, whatever that was, and she was pushing this too hard. Were the Yanks already putting her under pressure? Perhaps so, if they planned to expand their operations on Murdo and they wanted the run of the island. That would make a kind of sense. But why quite so heavy-handed? It wasn't the woman's style.

Hone knew one thing for sure. People did things for a reason. Especially people like Eugenie Webber. And Hone didn't like not knowing what that reason was. So he would find out. Information was the only currency he valued. Plus, Eugenie had attempted to bully him, which never went down well. His fingertips tapped the sand-filled lead pipe with the soldered ends he always kept in his riding coat. He liked to have it handy. You just never knew when you might need it. The problem with politicians and their operatives was they needed what he and his people could do. They needed those willing to work in the darkness, but that didn't mean to say they liked them. And they certainly hated the fact that in order to maintain deniability, they were required to let go of control. Eugenie Webber wasn't the first and wouldn't be the last person to think Hone and his people were her pawns to move around the board as she wanted. She would learn.

35

Murdo

North kept his back turned as Comfort slipped from the bed, and resisted the temptation to turn round as he heard the slide and rustle of fabric against skin as she pulled on her clothes. He had a sudden memory of Honor, the woman he'd loved, climbing out of a hospital bed and pulling on clothes, ready to do battle with monsters. A memory of how beautiful she was – how brave.

'I'd better head back up to the manse,' Comfort said. 'Enoch said he wanted a quick word about the celebration. But I'll see you in the refectory in about half an hour. Make sure you miss me, right!' She blew a kiss from the door.

It was bitterly cold in the cottage, the fire having died down to a few embers. He wanted coffee badly and food would be a good idea as well, since the last time he'd eaten had been in the bar on Harris that morning. Comfort appeared to have forgotten her promise to bring coffee and food back with her. But the need to see Mia was overpowering. And he could justify it. She'd been here five months and she should be an ally. She could save him time and energy if she would tell him what she knew about the place. He closed his eyes for a second to remember the geography of the settlement. As kids, they'd always hung out in the graveyard a few streets away from the children's home. The graveyard here was set behind the old church.

He checked his watch. It was half five in the evening. He could hear a shuffling and chat as islanders presumably headed back to their cottages or the accommodation block from whatever work they had been busy with. He didn't know their routines yet, but he imagined they would clean themselves up before they too headed for the refectory.

He put a call in to Fang. 'I think Enoch probably makes his broadcast from a place called the old church. Rowan – the young boy who was on the quayside – he told me that the YouTube and social media videos all go through the old church and that's where the offices are. As soon as I can, I'm going up there.'

'*Good to know. Remember: destroy the server, and take out the satellite link. If you can – kill the power to the place as well. And North, Granny Po and I are pulling out of the hotel. It won't make any difference to you. But we're going to spend a few days with Plug.*'

He frowned as he struggled to think why Fang and Granny Po would move out of their luxurious hotel to share a room over an undertaker's. But she had the answer ready before he could ask it. '*I'm doing some face-to-face interviews and Granny Po insists Plug goes with me.*' And North relaxed.

'Good move. He'll look after you.'

She snorted in derision then hung up. North had time to feel sorry for Plug.

Some kind of mist had blown in from the sea and North was glad of it as he made his way to the graveyard. He nodded at the few Narrow Yett followers he met and he could see the curiosity on their faces, but they didn't try to stop him or talk to him. He was trusting he had the look of a newcomer trying to get his bearings.

There was no gate. Only an opening in the stone walls that ran around it. Lichen covered the Celtic crosses that still stood. Other headstones were plainer. The inscriptions barely legible but for the odd letter. North wondered where Rowan's schoolteacher had ended up. He doubted there'd have been time to erect a stone for the man, barely time to dig a hole and lay him in it. North would change that. After what the man's library had done for Rowan, a grave marker seemed little enough to do for him in return.

What surprised him about the graveyard, though, were the graves whose mounds looked to be more recently dug and each marked with a small cairn of stones. The husband and wife who had leaped to their death? And three more. Had people died of old age here? Or sickness? The cult had been on the island for ten years. The chances were there'd been natural deaths.

He didn't hear her, only sensed her presence next to him. He turned. That round face with its remarkable patterning. The dark ringlets. The brown eyes – all he could see. Worth the boat trip. Worth the fight in the bar. Worth the bullet in his head to be standing with Mia again.

He didn't see the slap coming. Only felt the sting of her fingers.

'Mickey Xavier North, you complete and utter bastard.'

He rubbed his cheek. Even though she was beyond angry with him, he realized he was beaming at her. 'Hello, sunshine girl.'

'Don't give me that. And don't give me that grin either. It doesn't work on me, remember. What the hell are you doing here, North?'

North felt happier than he had felt since forever. She hadn't forgotten him. 'I wasn't sure if you recognized me. It's been a long time.'

Mia sneered. 'Maybe you hoped I wouldn't recognize you in case I kicked your ass.' She was raging. 'Imagine my surprise, though, when Comfort makes sure to tell me you're some rich party-boy. I mean, since when? Did you win the lottery? Did you rob a bank?'

He felt himself blushing. What else had Comfort told her? That there was a budding romance going on? By the look in Mia's eye – that, and more.

Mia shrugged. She could always read him. 'You're a grown-up. Sleep with whosoever you want to sleep with.'

'I haven't slept with her, Mia.' It was somehow important that she believed him.

Mia made a face as if to say she didn't care if he slept with every woman he met. 'A word to the wise, party-boy. Don't be fooled by the adorable fluffy kitten act. Comfort is as hard as they come. Enoch has his own praetorian guard of his most faithful followers and your girl, Comfort, she's part of the inner circle. She's a total piece of work and the only reason she's nice to me is because she thinks I'm her ticket to fame. I don't know what she wants from you, but there'll be something. I'm guessing – money. But maybe she wants you too?'

There was a beat.

'Are you just going to stand there? Taking this? You don't have anything to say to me?'

Hurt radiated from her and it hurt him in turn to see. Where to start?

And Mia's eyes narrowed as they met North's again. 'You were supposed to come back for me.'

They both knew what she was talking about. When North

went back to his mother after a spell in the children's home, he'd sworn to Mia he would come back and they would run away together. The home had been brutal and staffed with callous bullies and staff whose compassion had worn out years before. What made it bearable was the company of other children in the same predicament. In particular, Mia Barton. And he had meant it when he made his promise. He had it all worked out. He was going back to his mother, checking she was okay or as okay as she ever would be, and then returning for Mia. They'd figured they would squat in an empty house, earn money where they could, steal when they couldn't, educate themselves in libraries and look after each other. They figured if they picked the right house and kept their heads down, they could get away with it. The two of them were, after all, used to fending for themselves. This way, they would have each other.

'I couldn't come back, Mia.' He felt as if he was thirteen again.

She glared. 'You were a coward and you took the easy way out.'

His mouth was dry.

'Do you know how long I waited for you to come back? You forgot me.'

He shook his head. 'You're the least forgettable person I ever met.'

Tears brimmed in her eyes but she wouldn't let them fall, and it was like a hand wrapping around his heart and squeezing. But he had never been less than honest with Mia.

'I got home.' His voice was scratchy, but he persevered. 'My mum was supposed to have cleaned up her act, but she hadn't. She had some new guy in the house. Tony. He was...'

Brutal. Violent. Abusive. An addict. A predator. A thug. He

didn't need to spell it out. Mia knew the kind of man Tony was. She'd met the type herself often enough.

'He beat my mother and he beat me half dead one night.'

The mist swirled around them and Mia stood without moving a muscle. She didn't need colour and she didn't need exaggeration, only the facts. He'd been in and out of foster homes, and in and out of children's homes. But he'd loved his mother, and to the best of her meagre ability she had loved him too, but not as much as the euphoria heroin brought with it or the oblivion of booze. 'After I came to that morning, I found a hammer and waited for him. He went for my mother again – she was passed out on the couch – and I hit him. I wanted to kill him. Not just because I thought he'd kill her or because he'd hurt me, but because he deserved it.'

He'd had no regrets at the time and none after. Not for the act anyway. Only for the fact that he drew a straight line between his own incarceration and his mother's overdose six months later. And for Mia, the girl he left behind.

'They arrested me, tried me and locked me up. I couldn't come back for you.'

Her voice was subdued. 'You didn't change your mind?'

He remembered her warm hand in his when they walked along the street. The smile that lit up her face when she sat next to him at breakfast. Her courage facing down the bullies who called her a freak. How she'd lie on her bed with a library book and then tell him the story, acting out the parts.

'I didn't change my mind.'

How some nights when he lay in his bed in the young offender institution, he'd close his eyes and summon up the memory of her, that he'd go to sleep pretending she lay in the hollow of the bed curled up against him.

'But why didn't you write? Why didn't you call?'

'I did.' It had taken him a year. How would she feel about the fact he was a murderer? She'd hate him for it. Plug had been the one to say he had to try. That she was like them and she'd understand. He'd given him the phone card and North had called the children's home. But Mia was gone. And no, they couldn't help. He'd written a dozen letters and heard nothing. Then he had called over and over, till they hung up as soon as they heard his voice, and Plug said that was enough. North would have to wait till he got out to find her. But when he got out, too much time had passed. 'I kept calling...' It had been years and it felt like yesterday. '...but they said you weren't there any more. To leave you alone. That you had the chance to start over and I shouldn't spoil it.' And some hopeful part of him died inside.

She bit her lip.

'Where did you go, Mia?'

Hearing the fear in his voice, she brushed his hand with the back of hers. A gesture of reassurance she'd used when they were kids. She knew what he was frightened of – that wherever she'd ended up, it had been worse than what had gone before. Far worse. Because for children like them, it always got worse. 'I got fostered and they were good people. They ended up adopting me.' She sounded embarrassed now. Knowing he never had what she got to have. 'I got a home. But I didn't forget you. I looked for you too. I tried the care workers at the home and I even caught a bus to your house.' He felt his breath catch in his throat. 'But you'd gone. I guess your mum was dead by then? And you must have been in prison? I asked around but no one knew anything about either of you.' That didn't surprise him. If they had known, they wouldn't have talked about it to a stranger. 'It was okay,

North. My parents loved me. And it took a while for me to trust them. But I loved them too. They're the reason I'm here.'

He looked at her. 'I thought you were making a documentary.'

Even though they stood in the mist, she moved him to the shelter of the church and lowered her voice. 'They were scammed by these fraudsters. My parents signed over their house and the Narrow Yett started eviction proceedings almost immediately. I didn't know what they'd done. I was studying abroad at the time. They were frail, elderly, and ashamed they'd been so foolish. They saw no way out, so they killed themselves. They took pills, and when a neighbour found them they were holding hands.'

Fang's words came back to him. The daughter who took the Narrow Yett to court and lost. 'I went straight from uni into a job as a researcher in TV documentaries. I was careful to wipe away all trace of the "litigious daughter" and her social media – photographs especially, for obvious reasons. I went back to the name 'Barton'. Two years ago, I set up an indie production company. I've made three documentaries so far, as soft and squishy as you could hope for. It's taken me five years to get to the point where I had the kind of CV where I could make an approach.

'I told Enoch it was a great redemption story. He lost his wife, he lost himself, but he still found meaning in this life. He's a spiritual leader to these people. He's turning their lives around. I made out I admired him. A lot.'

Comfort had said she and Enoch were 'tight'. Had she had sex with Enoch? Again, Mia read the question in his eyes. Her chin went up. She did what she had to do and she'd do it again. North wasn't going to judge her – he used to kill people for a living. Occasionally, he still did.

'Enoch is a narcissist and a sociopath. He wants to be famous. That's what all of this is really. One giant "Look at me!". He wants to use his fame as a recruiting tool. Recruit more members he can control. Earn more money. Enoch thinks I'd do anything to get this documentary made. He doesn't think I'm a journalist. He thinks I'm in the entertainment business. That I'm young enough to be controlled and that I'm in talks with Netflix.'

'Is that true?'

'I say what he wants to hear. I didn't know what I'd find when I came out here. And I have to be careful they don't catch on. It's taken time but I think I've nearly got everything I need now. Mutilations. Sexual abuse. Child neglect. Violence, intimidation, brainwashing, financial fraud. They think it's normal. In interviews they don't even realize what they're telling me. I'm going to mop up some footage from the tenth anniversary party and I'm out of here. There's a kid. The one on the quayside—'

She meant Rowan.

'I met him.'

'He won't leave without his sister.'

North knew Rowan's plight would have been killing her. 'I was hoping I could persuade him. The authorities would have to take my footage seriously and come out here. I already knew they were interested in this place...'

'What do you mean? How did you know that?'

She hesitated. 'They approached me. Told me they knew about my parents. They helped me get on the island.'

North took a moment to think through what she had just said. Why hadn't Hone mentioned that?

'Anyway, North, none of this answers why you're here. Do you expect me to believe this is a coincidence? That you just happened to get in a fight and throw yourself on a boat to the island where I'm living? How did you find out where I was?'

He felt the pulse in his blood quicken. The connection between them had been brief and intense as kids. As adults, it had a different quality to it. The friendship flickering and smouldering into an unmistakeable sexual tension.

She thought he was here for her. She had no idea he might be connected to the security services. Why would she? Something flared in her brown eyes. A spark of what might have been joy. They softened as they looked at him. 'You came all this way for me?'

'In a way,' he said. And it was true. If Mia wasn't on Murdo, he wouldn't have come. He'd have resisted Hone's blandishments this time.

'In a way?' That focus again. He told her all of it – Gaffney. Tower Bridge. The imminent call to violence. And she scrabbled at her phone and scrolled through what he presumed was her filming. She held it out to him. A queue of islanders standing on the rocky shore. A figure thigh deep in the sea. The picture zoomed in on the figure. Jonny Gaffney was holding on to a young woman. He pushed her backwards into the sea, still holding her. Then brought her up, gasping for air. Down again and up again. Down and up, each time holding her longer under the water. 'They bang on about computer simulations all the time, but Gaffney saw himself as John the Baptist,' Mia said. 'The prophet who went before Jesus and prepared the way. He also ended up with his head cut off, and from what you saw, Gaffney tried his best to do the same thing.'

'Which leaves Enoch as our very own "Cyber Messiah",' North agreed.

'I heard Gaffney used to be a nice guy but when I first met him I thought he was really disturbed. And in the five months

I've been on Murdo, he only seemed to get worse. But if people here know what he just did down in London, that explains why the mood down there is off. Everyone's jumpy. Enoch's apparently locked himself away and is having some kind of existential crisis. I can't get in to see him. But from what you say, it might not be a crisis so much as preparing himself for battle.'

North spared a thought for Comfort. Had she been allowed in to see Enoch? What would she find when she got up to the manse?

'I have to stop the broadcast,' he said. 'If I don't, innocent people are going to die. And I have to kill Enoch. You don't have to be part of any of that, Mia.' Hone had inveigled Mia on to this island and now she was in danger. He wasn't dragging her further into whatever kind of mess this was.

She stretched out her hand and gripped his forearm and he did the same to her. Their handshake when they were kids. He had never kissed her. Not once. They had been friends.

'I know my way round this place, and I know what makes these people tick. And Enoch killed my parents as surely as if he had smothered them in their sleep. I'm all in, North.'

He should have been pleased. Mia was going to help him, just as Hone had wanted. And part of him was. But Mia had been safe as long as Enoch thought she was doing his bidding. Drummond already resented her presence. If anyone found out she wasn't here to make Enoch famous, but to destroy him and his cult along with him, Mia would never make it off the island alive. This was a bad place full of bad people. And Mia knew all of their secrets. North's eyes went to the graveyard. There was plenty of space here for unmarked graves – plenty of room for Mia.

36

The mist was starting to clear and they decided it would be better if they weren't seen together.

'I don't want to leave you,' she said. And he felt himself start to smile again.

'Behave like you always do, Mia. You can't have them thinking anything is wrong. Think about how we get to Enoch.'

She gave a small nod as if to say she saw the sense in what he said, but didn't like it. And he watched her go. Mia Barton, all grown up and even more impressive than she had been as a kid, which he hadn't thought possible.

Now he had to get into the tech hub and kill the power to the place before anyone saw him.

The former church acted as a porch to the sprawling brutalist structure behind it. Clad in huge sheets of what looked like rusting metal, held up by massive pillars, it was stark and brooding, without even a window to its name. A huge satellite dish sat atop the roof. Even getting up there was going to be a problem. But he had to try. As for the best way to knock out the power, he had to recce the place first or he'd be going in blind. The problem was if he went in the old church door, he risked someone inside challenging him. But if he went in via the door at the back, it would be obvious if someone caught him that he was nosing around. He hesitated for a second, his hand on the handle of the rear door. Should he wait for night?

He was thinking so hard about not getting caught that he almost missed the dull murmur of conversation on the other side of what turned out to be a heavy steel fire door. There was nowhere for him to hide. Grabbing a rock, he wedged it under the bottom of the door and pressed his full weight against it. There were muffled protests on the other side. How many were there? He could hold it, but only for so long.

'*Is it locked again?*' He heard swearing.

There was a last attempt to push the door open but North held on to the handle and dug in.

There was fresh cursing and he pressed his ear against the metal. They were giving up and moving off. He gave it a minute, removed the rock and opened the door as narrowly as he could. He slid in.

It took a second to adjust to the gloom. The door opened into a narrow corridor that smelled hot and musty. Step by cautious step he advanced, pressing himself against the walls.

Although the cottage was lit by a paraffin lamp, along with its neighbours, it had been obvious on his arrival that the Narrow Yett had electricity. Not least because there had been a dozen fuel tanks at the quayside and a small power plant. When he'd arrived there'd also been electric lights coming from the manse and from the refectory. North wondered if it was easier to keep the cottages free from electricity or if their occupants enjoyed the authenticity of paraffin lamps and turf fires.

Inside, it was a far cry from paraffin and turf fires. North peered through a huge plate glass window overlooking what appeared to be the main atrium. At the far end, a huge screen played muted video of a man North presumed to be Enoch.

The nose was narrow and the mouth wide. The shoulder-length hair fell from a widow's peak over a high forehead and was shot with silver, matching the stubble on the jaw. Leaning into the camera with brown unblinking eyes, Enoch appeared to be talking without hesitation, occasionally bringing his hands together and closing his eyes as if in prayer. The prayer never seemed to last long. North figured he enjoyed one-to-one time with the camera too much. Behind him, pictures of environmental disasters played on a loop.

The sound went up, as if somewhere someone had decided to check the quality of the recording.

'I've heard people describe the Narrow Yett as a cult.'

No kidding, North thought.

'I want to tell you that's a word used by the powerful to disempower those who refuse to accept the orthodoxy of the moment. Swap out the word "technology" for "God". Swap out the idea of moving to a new "simulation" for the word "Paradise" or "Heaven". Does that sound like a cult to you?'

Yes, North thought.

'Humanity has always believed we can move between one state and another, between this life and another life in the hereafter. Do you know why? Because deep down within us, we realize we are only so much code. That is what DNA is, after all – so much code. Deep down, there are plenty of us who realize that this Life is only one simulation of many. And that we can – if we are fortunate – move to another simulation.'

Enoch's voice was warm. The accent pure Scotland. And his gaze appeared warm, albeit unblinking. But this was a leader with two million followers, North reminded himself. He was always going to have something about him. Enough charisma to persuade otherwise sensible people to abandon their homes and their loved ones and pass over all their money.

Enoch carried on. 'We in the Narrow Yett are prepared to think differently. We're proud of the fact we are a modern social movement with a spiritual and ecological mission. We're proud of our charitable giving. Do you know how many people have accessed the Narrow Yett teachings via social media? Ten million people. Does that sound like a cult to you?'

Still yes, North thought.

'And tomorrow at noon, I'm going to be making an important announcement that will affect each and every one of us. So keep checking in. Until then...'

The sound dipped again. This was it. Enoch was building interest in what would be his call to arms, North realized. Did the members of the cult on Murdo know what was coming? The tech guys in the atrium didn't seem excited at the prospect, but he guessed they had seen the broadcast before. Plus, the vast majority of them were Drummond's people, of course. Between the window and the screen, more than a dozen heads were bowed over monitors, some of them running a laptop and a monitor. All of them ignoring Enoch's sermonizing.

North was too far away to be able to make out the detail on the monitors, but this had to be where the money happened. To the side of the central screen, underneath a sign that read: 'Making this reality a better one for all', a display showed a clock, a world map, and the donations from Enoch's YouTube channel ratcheting up. In a five-hour period, the Narrow Yett had earned just short of £1,000. Not bad. North did the maths in his head. This was a global operation. There were 8,760 hours in a year and that added up to a cool 1.752 million pounds. And he had to guess that occasionally Enoch laid on fundraising drives and called on his audience to dig deep. Perhaps that ramped up the total to 2 million or even 2.5 million pounds? And that was without any online courses or the 'spiritual' retreats on Murdo for those who wanted to experience the Yett in real life – or as close to real life as they believed they could get. It also didn't take into account members like Gaffney or Mia's parents who signed over the family home and had it sold from under them. It was serious money. The Narrow Yett was not so much a cult as a money-making machine, North thought. Especially when you took into account the fact that Drummond was washing his own cash through it. Drummond had no idea what his old friend was planning, North realized. He wouldn't sabotage an operation like this. There was too much money at stake.

He needed to find the main server and bring these computers to a grinding halt. Then get up to the roof. Step by step, he edged back towards the shadows, which is when he felt a hand on his shoulder.

37

London

Fangfang Yu was on the move. And if Fang was on the move, so too was Granny Po.

She had to assume the CIA had someone watching her. It couldn't be electronic within the suite because she swept it regularly and it was clean. But who knows how many of the maids and waiting staff were reporting back to Roxie? Which gave her no flexibility.

Her original thought was to leave Granny Po in comfort at the Empire, but Granny Po had different ideas. And as Fang had learned a long time ago, there was no point in resistance when it came to her grandmother.

She tapped out a code on her phone and imagined the security staff frantic downstairs as they attempted to get the CCTV, which had just turned to black, up and running again. *System error*, the screens read. *Attempting restart in 60 seconds, 59. 58. 57...*

She crooked her arm and her granny slid her hand through it, and together they headed down the corridor to the service elevator, which rattled and shook as it descended.

Would Roxie have someone on the back door as well as in the hotel? Fang thought she would. They stepped out into the service corridor and headed for the third door on the left.

Fang shut the door behind them. She searched the shelves for the dark aprons she wanted and handed one to Granny Po and wrapped one around herself. The aprons were huge on the two of them.

This was it. If Roxie's people spotted them now, Roxie would sweep them both up and Fang would be on a plane before most people had put the kettle on for a cup of tea.

She nodded at Granny Po and the old lady followed her back out the door and along the corridor.

It was all in the timing.

A huge bespectacled man with a carefully groomed beard and wearing a dark apron was approaching the doorway, clutching a massive display of slightly wilted flowers and luxurious foliage in a white Grecian urn. Three young porters trailed behind him, each one all but hidden by their own enormous displays of purple delphiniums, black tulips and white roses. Granny Po seized hold of two huge armfuls of flowers from the first porter. 'We're with him,' Fang said, gesturing to the departing florist, and the two remaining porters dumped both loads on her with barely suppressed relief. The procession of dying flower displays curled its way through the service corridors and out the back door to the waiting van.

Fang held her breath. If they were lucky, the only thing any watcher would be able to see right now were feet. If they were unlucky, the watcher would realize who the feet belonged to. The bespectacled man had already thrown open the back doors to the van. He seemed oblivious to who was helping. Granny Po heaved herself up into the van and Fang followed. Inside, it stank of old flower water and roses. Fang eased the doors closed then pulled the displays of green leafy branches and only just fading roses up in front of them to create a floral wall against the world.

The door to the van slammed. There was the noise of boots on the tarmac and the driver of the florist van started the engine. Granny Po half fell against Fang as the van pulled out into the early morning London traffic.

The van rattled and coughed its way through, stopping and starting, bumping over what felt like cobbles, and by the time it drew to a halt, both Fang and Granny Po were nauseous and covered in faded blossoms.

There was a metallic clunk and a gust of cold air as someone threw open the doors to the van.

Fang peered out through the flowers.

In the doorway, the huge bespectacled man took hold of the flowers and dragged them to the front of the van. The ceramic base of a vase screeched against the floor.

'Let's be having you stowaways then,' Plug said, peeling off the beard to reveal a wide grin across his slab-like face.

Granny Po scampered to the front of the van and Plug scooped her into his arms and set her down on her feet as if she were made of precious porcelain that might shatter if a wind blew on her. The old lady patted his arm in thanks, gazing up worshipfully, and giggled as if she were the teenager rather than Fang. Plug slammed the door again, plunging Fang back into darkness.

'Unbelievable,' she grumbled. 'I work with idiots,' she said as she crawled out through the debris, pushed open the door, and jumped onto the concrete floor of the East End undertaker's.

Plug put his finger and thumb in his mouth and whistled. 'Oi. I don't pay you to count your spots,' he called out, and two young lads emerged from the shadows with buckets of soapy water and scrapers for the florist company's decals. An industrial paint gun loaded with grey paint was already lined up on a shelf, along with new registration plates and a cross-head screwdriver.

The real Florian arrived at the Empire as the fake Florian left. If he was surprised that the hotel had been cleared of flowers past their best, he never mentioned it to the duty manager. If they had felt the need to remove them before his arrival, it reflected on him. Best let sleeping dogs lie, he decided. Instead, he arranged the new displays of peony roses and amaryllis he had arrived with, and pocketed the 'In Loving Memory' card from an ornate display of irises and white chrysanthemums from the Georgian side table in reception. The contract here was his and he felt a flutter of nerves that another florist might be trying to encroach on his territory. Still, it looked rather splendid, so he contented himself with breaking one or two of the iris stems and let it stand. Tomorrow, he decided, he would replace it with a display of *Phalaenopsis* orchid plants in a round green-glass vase, so beautiful that hotel guests would weep at the sight.

'What about the CCTV cameras en route?' Plug asked with a frown.

Fang nodded. 'Taken care of.' Fang had connections and her connections had the ability to down CCTV cameras across London for an hour.

'You're safe now, kiddo. They won't come for you here. Let them try.' Plug's ugly face was savage. She'd confessed about Roxie and he hadn't been happy.

Fang considered his statement as they made their way through the door into the rear of the undertaker's premises and as she ploughed her way through the chilli and rice and drank the orange juice that appeared in front of her.

She knew, even if he didn't, that it would only be a matter

of time before the CIA rooted out Plug's name as a known associate of Michael North, if they hadn't already. At least she'd managed to warn North about Drummond and the money laundering. The question now was, would she have enough time to discover what was really going on at Benson – before Roxie came for her. Because when that happened, Plug would defend her to the death, and that was a problem. The situation Fang found herself in was dire, and she worked for MI5, which appeared to count for less than nothing. Plug didn't even have that much protection. She couldn't allow him to pay the price for her stupidity.

The clock was ticking. They had just under eighteen hours till Enoch's announcement.

Bearing in mind that rooting around in Benson's stacks had brought the wrath of the Department of Defense down on her in the form of CIA maven Roxie Powell, when it came to Benson, Fang was watching for anything out of kilter, however tangential. She'd already been through their financials, taking infinite care not to trip any kind of alarm. She'd already gutted the CVs and taxation records of their board members and management team. She'd already read every word there was to read about business tycoon Mrs Elaine Benson, otherwise known as 'the Bloody Widow', and admired the photographs of her chiselled son, Teddy Benson. And she was currently crawling over any employee ever mentioned in any scientific and press report. One thing caught her eye: the recent death in the Scottish Highlands of satellite engineer Dr Lawrence Sampson in a car crash. A road traffic incident involving just the one car. Dr Sampson, according to reports, had fallen asleep at the wheel and driven off the road.

Yes, Roxie had made it crystal clear that Fang was to keep out of Benson's business. Which made it all but certain she was going to turn them upside down and shake them till they rattled. Fang popped in a piece of raspberry-flavoured bubblegum as she leaned back against her seat. She chewed till the synthetic fruit flavour was gone and then she blew the biggest blue bubble she had ever blown. Blew it so big and so thin-skinned that it could do nothing but pop. The noise reminded her of her cheekbone hitting the glass window of the London bus. She wondered if Dr Laurie Sampson had family in the UK and, if so, what they'd been told about the crash.

38

Murdo

In the tech hub, North pivoted, and as he did, he brought up his arm, ready to thrust his elbow into his opponent's throat. He stopped – a moment before smashing Comfort's larynx to pulp.

Comfort's blue eyes were wide in terror. Her mouth opened to shriek her protest, but North clamped his hand over it and dragged her into the lee of a steel column. From a staircase along the corridor, there was the sound of steps and of conversation. He already knew it was too late. They were too obvious and they had no time to get out of there. Instead, he pulled Comfort to him, wrapped his arms around her and kissed her. She squirmed against him, trying to pull away, but he kept hold.

Drummond turned left instead of right. He was walking away from them. His head down, listening to one of the techies. The younger man appeared to be busy explaining something, but some primitive warning system must have made Drummond turn around and look back down the corridor.

'What the feck are you two doing in here?' Drummond bellowed.

North glanced up, as if annoyed at the interruption. Rage

distorted Drummond's face, turning it into something brutish and primitive.

North broke apart from Comfort, moving in front of her to keep her away from the other man's temper. Drummond moved closer, and his hand went to his belt. He must have a knife on him, North thought. And he wished he did too.

Coming out from behind North, Comfort stepped between the men, raising her hands in a submissive surrender position, the palms facing Drummond. 'It's cool. I was giving North the official tour before supper,' she said. 'Take it up with Enoch if you have a problem.'

Drummond gave every appearance of fury. He narrowed his eyes as he glowered first at Comfort and then at North, as if considering the physical hurt he could do. Then his phone bleeped and he glanced down at it, then back up at them.

There was a beat as both men considered whether this was the place and the time to take each other on – before the phone bleeped again, and Drummond moved off the way he had come, the young techie almost running to keep up. 'Feck the feck off, North. You too, Pussycat.'

'Oh my God.' Comfort was giggling, but it was from nerves rather than anything else. 'What were you thinking? Why are you in here?'

He kept it casual, shrugging. Drummond and the young techie had taken a right off the corridor. Was there any way to find and destroy the mainframe server now? To make it up to the roof and the satellite dish? Not without knocking Comfort unconscious. Not with Drummond in the building. He'd have to come back tonight.

'You said see you at supper. I thought maybe this was the dining hall.' Comfort's nose wrinkled as if she could smell the lie on him. She tugged at him to get him moving. They were following the same route as Drummond. She took it cautiously, as if nervous that Drummond could emerge from one of the rooms off the corridor at any point.

He'd have said more, but his eye snagged on another screen through an open doorway. It looked as if it was some kind of edit suite, although one wall was made up of mobile phones, each with a small label under them. They had to belong to the cult members, he realized. They must have given up their phones when they arrived. Mia's face was on the monitor. She was standing outside and speaking directly into the camera. The sound was muted. Comfort tugged on his hand, but not with any great conviction. She wanted to watch too. She just didn't want Drummond to realize they were behind him.

Drummond waved a hand at someone sitting at a desk. He must have told whoever it was to bring up the sound.

'... a charismatic leader who offers a genuine alternative to the troubles of today's reality...' Mia was saying. 'Many of his followers believe the Narrow Yett has transformed their lives for the better. To its supporters, the Narrow Yett is a new religion – a cyber religion – offering an ecologically viable, socially minded belief system with a sound moral code that offers a genuine alternative to traditional religions. The Narrow Yett moves faith out of an archaic church setting into both virtual and real life. It's wildly popular, it's real and it's here, whether this is reality or indeed a simulation.'

Mia wasn't kidding then, when she said it was propaganda.

The pictures cut to a sunny day. Comfort was smiling. She had tied the blue hair back into a ponytail. 'I genuinely believe I'd be dead by now if it hadn't been...'

North glanced down at her. Her lips were moving in sync with the recording. '... for Enoch and the Narrow Yett. I never had a family before. I do now.'

Drummond sat down as if he was too bored to stay standing. North got the impression he had had it with paeans of praise directed at Enoch. 'The rest of it...?'

A burly heavy with ginger sideburns that ran down to his jawline gave Drummond a smug grin. 'We should be paying her, boss. We should persuade her to let us have the rights for the channel. We could match what some streaming service would offer. I can't even see them being that interested.'

'You can't see them being interested?'

The ginger heavy shook his head. 'Nah. Her stuff is a huge yawn. She's not getting anything spicy.'

Drummond scratched at his skin under the black beard. 'Five months here and all she comes up with is this? She doesn't strike me as someone who'd miss the fact this place is a freak show.' And North felt the hairs on the back of his neck stand to attention.

Comfort was dragging on his arm again and North allowed himself to be pulled away. A doorway had been knocked into the stonework, and they emerged from the corridor that wound around the tech hub into the shell of the empty church. Light fell through the clear glass windows of the old church, casting bright stripes of white along the stone floor. Comfort shivered. The tech hub had been warmed by the machines, but there was no heating that he could see in the church, only a scattering of leather chesterfields and coffee tables. A dartboard hung on the wall and a small fridge full of beer hummed in an alcove where there had once been a baptismal font. It looked as if working in the tech centre came with perks.

Comfort gave a small skip as she pulled him past the sofas and over to the wooden door.

'Let's get supper,' Comfort said. 'You don't have to go nosing round on your own. There are no secrets here.' She flashed him a smile meant to disarm him, and he knew she was lying. The place was full of secrets.

The main community hall building was constructed out of the same rusting iron slabs as the tech hub, though the front facade was cloaked in silvering untreated timber. Enoch had invested serious money in the design of his buildings on Murdo. Even so, from where North was standing, the refectory looked like a rusting prison ship.

His expectations were low as they entered the hall. Instead of an interior of darkness, however, the community hall was filled with light courtesy of the floor-to-ceiling windows overlooking the sea on two sides, and it was heaving.

There was a moment of silence as he entered. He could understand that. The island was in the middle of nowhere, and they hadn't been expecting a visitor. It took a second before the normal buzz and swoop of conversation resumed, but North sensed the disapproval in the eyes that stayed on him. They must know the story by now – that he'd saved Mac, and in saving Mac had saved the girls. But then this wasn't the kind of place anyone could slide into and slide out of without drawing attention to themselves. Rescuer or not, though, they didn't welcome his presence.

Servers moved between the tables carrying trays and jugs. The hum of voices and the smell of spice and the warm fug of bodies gathered together. He looked for Mia in the crowd but couldn't see her. She'd said she'd be here. Where was she? He had to warn her that Drummond was on to her.

And where was Enoch? The reason for sending him here. Killing Enoch in front of three hundred witnesses wouldn't be ideal. It would create a huge stink and he'd probably end up dead himself. But it was an option. Surely, the man needed to eat, even if he was having an existential crisis, as Mia claimed.

Comfort steered him to a long trestle table and pushed him down onto the bench seat. She leaned over him, her breast grazing his arm, and passed him an earthenware bowl and a chipped mug. Almost immediately, a ladle appeared behind him with a dollop of brown rice and then another – this one a loose slop of curry-scented lentils, chopped tinned tomatoes and skinny translucent slices of onion. A huge steel teapot was passed up towards them. The tea was lukewarm and had barely seen a teabag, but Comfort didn't appear to notice. She picked up her spoon and started to eat the lentils. He stared down at the slop. To call it stew would be an insult to stews everywhere. It was a grey and lumpy soup. But he was ravenous, and he'd had worse. He ate it.

'Where's the famous Enoch?' he asked. An innocent enough question. Comfort had mentioned him. Any visitor would be curious.

Comfort looked vague. 'He's dealing with something.'

Dealing with Gaffney's actions? Planning the best way to bring violence to the streets? Because North could see that was something that would keep a person busy. Would Enoch have guards up at the manse? Mia had mentioned he had his own version of a praetorian guard. His most faithful followers. Because if Enoch wouldn't come to him, he would have to go to Enoch.

'Tell me about yourself,' Comfort said, and he wondered if she was changing the subject.

North regarded her evenly. He couldn't help but like her. He'd kissed her in the tech hub. He was sending out signals saying he was interested, so it was natural she'd want to get to know him better. Then again, Mia had warned him not to be taken in by Comfort – that she was hard and manipulative. That Comfort was in Enoch's inner circle of true believers. And in his experience, Mia was an astute judge of character.

What had this afternoon really been about? Comfort had been naked in his bed an hour ago. Was it possible that she'd been told to sleep with him? That Enoch had guessed who or what North was? That he'd know Gaffney's atrocity would trigger official enquiries? He'd taken her at face value. It was possible that he shouldn't have.

'Not much to tell.'

Would she push?

She fluttered her eyelashes.

Yes, she would push. The term 'moron-person' did laps in his brain. The girl might well like him, but she was still engaged in finding out his history. 'I'm sure you're wrong. Speak your truth, Michael North. What's your story?'

He had a humdinger of a story – neglected, beaten, murderous, imprisoned, wounded, saved, recruited. And now here to break apart the Narrow Yett.

'My "truth"?' he said. This was it. They were suspicious of him, but not sure. Certainly, anything he said would be checked and double-checked. Hone had said it would bear scrutiny. All he had to do was deliver it with authority. Forget how far distant it was from his own miserable childhood, from serving time for murder, from the army, from being catastrophically injured, from killing for a living.

He took a breath. 'Dad's Andrew and works in the City. Big yawn. Makes a shedload though. Mum's Fiona. She's an interior designer. Obsessed with cushions. One sister, Mandy. Love them all to bits but have to say home's a yawn. Good school. Eton, actually, although the bastards kicked me out.

Still keep up with a lot of the guys though. Partied a bit too hard at uni. Gramps kicked the bucket and left me some dosh so took off travelling. Love Thailand. Unbelievable people. So authentic, you know. Want to go back. Really fancy India. Hear great things about Goa.' So smug. So entitled. He wanted to punch himself in the face.

When they checked it out, they'd find the fictionalized Michael North also had a diagnosis of ADHD and a local press report of a driving ban courtesy of being over the limit.

But no one would fess all within a few hours of meeting someone. 'Posh boy' North had to have a few things left to share later. It just made the rest of it more plausible.

If Comfort didn't buy it, he would have a problem. A big problem. Three hundred problems.

The blue eyes didn't waver.

He'd rejected her advances. And she'd just caught him sniffing around the tech hub. She was in Enoch's inner circle.

Had she bought it?

'And what do you do for a job?' She'd bought it.

He waved a hand dismissively. 'I haven't found "my thing" yet. You know.' Did young privileged men really talk this way? He had no idea. But from the supportive nodding, up and down – they did.

'But what do you do for money?' she said, her eyes wide and innocent.

North leaned in towards her, keeping his voice low. 'All sorts.' He tapped his nose as if it were a joke. In North's experience, the only people who considered money a joke had way too much of it.

Comfort nodded as if she knew exactly what he meant. Her big blue eyes were shining. But she was harder to read than he had first thought.

A wooden bowl of dark bread appeared, along with a block of cheese, and he helped himself. It was bitter and tangy, sheep's cheese, he thought, but after the lentil slop it tasted good. Comfort's gaze switched to the cheese. He pushed it along to her, but she shook her head.

'It's good,' he said.

Again, she shook her head. 'I haven't earned it.' She chased what was left of the stew around the bowl and licked the spoon. She didn't seem to think what she was saying was in any way abnormal. 'We need to keep our bodies pure so that we're always ready.'

'Ready for what?'

'To go through the yett that links this simulation with the next. It's important not to corrupt our code. But guests eat differently to the faithful. The Father says it's important to be disciplined and not give in to our baser appetites. We earn privileges through reflection and study and our work.'

He looked at her, and then around the hall. Many of the faithful were skinny. Most were devouring the slop. Only one or two had the privilege of bread and cheese in front of them. And a few sat with bony hands folded, sipping water or clutching mugs, their faces a picture of misery and longing. At least three of the men – maybe more, he couldn't tell – wore pouches around their necks. He had to hope they didn't hold what he suspected they held. North's childhood had been one of abuse and violence. But he'd got past it. He'd suffered a traumatic injury when he took the bullet to his skull, but he'd found a way back. He'd lost the woman he loved most in this world, but had refused to let it define him. He could die at any second, but chose to embrace life rather than fear death. What terrible things had the people in this hall been through that they had washed up here, on a rock in the middle of an ocean, in a collective delusion that this community was a good thing? That it would save them from reality?

How long had Comfort been on the island? He'd thought she was an innocent, but Mia said she was a piece of work. Perhaps the blue-haired girl simply didn't know which way was up any more? He carved a piece of bread and a chunk of cheese and passed it to the girl next to him. 'We nearly died in that storm. You did good out there – you can eat a piece of cheese.' Comfort looked first at him, then around to see if she was being watched, and eased off a morsel of the cheese. Whether she would have eaten more, he didn't know, because there was a flurry of activity as the door behind them went and Drummond entered with his men.

Unless he stood, he had no sight line to Drummond, surrounded as he was by his cronies. But North didn't need to see the guy to feel the darkening in the mood. The gangster didn't have many friends here either.

Comfort nudged him as Drummond started walking towards them. His heavies and tech cronies staring after him as they headed for their own long table in front of the picture window. A table already loaded with food and bottles of beer.

North stood. Had Drummond thought about his presence in the tech hub and decided it was suspicious? Was he about to throw him off the island after all? Which would be a good trick since there was no way on or off until Mac rocked up again. Somehow, he didn't think that would stop Drummond. He balled his fists in readiness.

'Sit the feck down, North,' Drummond barked. It was an order rather than an invitation. North considered what would happen if he didn't. They'd fight. He would win or he would lose. In either eventuality, the mission was over.

He sat. But as he did, he felt Comfort's touch on his back.

'Catch you later.' She was already moving off and leaving him to it.

The air filled with the fragrance of strong coffee as Rowan's mother, Lisette, appeared with a pot and two beakers. Drummond might be a bastard, but North wasn't turning down good coffee.

Unlike Comfort, Drummond didn't seem all that keen on keeping his code pure. A platter of sliced beef and ham appeared, as well as steaming potatoes and brown bread. There were hard-boiled eggs, which had already been peeled, soused herrings and thick slices of tomato and red onion in some kind of vinaigrette. Whatever the Narrow Yett was, it wasn't a democracy, North decided, staring at the food. Drummond mistook his look for hunger. He cut him a slice of the bread, piled it up with beef and pushed it across.

'Eat,' he said. 'The slop they serve here doesn't do much for the likes of you and me.'

North remembered Rowan frying his skinny fish in his cliffside cave lined with the bones of a whale, and hungry though he still was, his appetite left him. Drummond didn't appear to notice.

They sat in silence. Drummond apparently relaxed and North determined not to be the first one to break it. Finally, his meal over, Drummond mopped his mouth with a napkin and spoke. 'I hear you met an associate of mine.'

'News to me.' Had someone seen him and Mia in the graveyard? He should never have exposed her to danger like that.

'Little ratty Yank. Burns. Aviators. Toothpick.' North hadn't seen that one coming. 'Kincaid just called me. Said you beat up his men over on Harris. Said he was hoping you'd be "sleeping with the fishes" by now. Said to watch out for you in case you made it over, and I'd be doing him a favour if I erased you from existence.'

'The guy bears a grudge.'

Drummond shrugged. 'I like that in a person.'

'He's a friend?'

'He and I share an island with a bunch of tree huggers who don't have any trees to hug. We do a bit of business together. Occasionally, we have a game of cards. Sometimes I bring him over a loaf of banana bread. I like to bake.'

North raised his eyebrows.

There was a flash of white teeth in the black beard. 'It relaxes me.' Drummond could be charming when he wanted, North realized.

The gangster leaned in closer. 'So, you picked a fight with Kincaid's men and then you beat the crap out of them. You jumped in the sea after that old fart Mac. Despite how perishing it is, you get him back to the boat. And you've got the balls to go toe to toe with me on the quayside. And I'm supposed to believe you're over here chasing a piece of skirt? I don't think so.'

North sat back on the bench. Drummond wasn't buying it.

'I'm a reasonable man, so I'm giving you the opportunity to tell me what you're up to. We got off on the wrong foot. That little runt winds me up, and he had my kid. Rowan's weak like his mother. I've tried to toughen him up is all.'

North thought again about Rowan, cold and hungry and living in a cave for the past three years. He remembered the men in his own childhood who'd believed they were doing him a favour when they doled out a beating.

Then again. If he let Drummond know the game was up when it came to the Narrow Yett, he'd have allies. Enoch and Drummond were old friends, but Rowan had said Drummond was getting fed up. That he was close to shutting his operation down here. It was true that the man was a gangster, but Hone wasn't interested in criminals. If Drummond cooperated, North would have two dozen armed men behind him. Enough to take apart the tech hub and put Enoch under lock and key. He wouldn't even need to kill him. With Drummond's help, Enoch would be neutralized and the other members of the cult could be kept in containment till reinforcements were sent in from the mainland.

It was the right play. Drummond was a man used to making strategic decisions. He might not even have to serve time for the money laundering, because Hone would make a trade. Not least because Drummond would be able to provide intel on weapon and drug deals with links to terrorists all over the world.

The crime lord watched him.

He knew exactly how it would feel when Drummond's spade-like hands hit a young boy round the head. The stumble and the daze and the pain of it. How the force would travel down the spine till it felt like every bone was ringing with it. If working with Drummond was the right play, North wanted to play it all wrong.

A dark cloud settled back across Drummond's face.

'Okay, play it your way, North. So long as we're clear about where we stand. Remember, I reached out a hand of friendship. You just weren't interested.'

He drained his mug of coffee and pushed his chair out from the table. He raised a hand and across the hall his cronies stood as one man. 'Everyone's expected to make a contribution round here. Tonight, you can help out on the building site. We've got a construction gang working late building the schoolhouse. And tomorrow you can go out on the harvest. Show you what island life is really like. Up for it?'

What could he say?

He could hardly refuse. Even though they both knew the fact Drummond had plans for him was never going to be a good thing.

39

It was barely first light as the truck bumped across the scrubland of the island, and North felt the eyes of the other passengers drilling into him. The construction work through the night had been ten hours of hard manual labour in darkness lit only by paraffin lamps. The men he worked alongside seemed surly and resentful that they were working on the site after supper. And North became convinced that Drummond had called the gang together with the single task of occupying him. He'd barely been back in the cottage twenty minutes when there was a knock on the door, and he'd been escorted to the truck. Since he'd taken a bullet to his head, he suffered from acute insomnia. So the lack of sleep didn't bother him, although he presumed that it was meant to both weaken and disorientate him. The bigger problem was that the labouring had left him no opportunity for a return to the tech hub to destroy the server and the satellite dish.

As the sun rose, pink and gold streaked the blue sky, and it became clear that they were a motley bunch. A couple of malnourished men in beanie hats, one green and one yellow, both with pouches around their necks, and Joy, who didn't meet his eye and was doing a good job of pretending she didn't know him. There were also two bruisers in the front, and from the scowls they both gave him, North was guessing these were part of Drummond's crew. He could hear them muttering and from what he could make out, the conversation was about Gaffney.

Both of the bruisers appeared shocked.

There was a divide in the community, North could see for himself, between Drummond's thugs and the Narrow Yett members. Whatever they were doing on Murdo was simply a job for Drummond's people. A way to make a living. Not a cause and not a religion.

A Jeep followed on behind. In the passenger window he could make out Drummond's grinning face.

He attempted to take his bearings. As far as he could tell by the position of the sun and what he knew of the topography of the island, they were heading north-easterly, which surprised him. The only thing up there as far as he could remember were vertiginous cliff faces.

Nobody had told him what it was that Drummond had in mind. Last night, he'd said something about a 'harvest'.

'What are we doing?' he asked Joy in a low voice. She frowned at him.

'Didn't you volunteer? I thought with you being an action man, you might have volunteered.'

Never volunteer, he remembered from his army days. It could get you killed.

He remembered what else Drummond had told him. 'No. I guess I'm being useful.' He glanced down at the flatbed of the truck. Long wooden poles with rope nooses.

'Back on the mainland...' She said it as if the mainland was as far away as Mars. '...do you climb?'

He shook his head. He had climbed on military exercises in the past, to go up or go down. It seemed pointless otherwise. And he had made a pretty poor fist of it yesterday with Rowan.

The beanie men looked at each other and then at the woman. They seemed concerned. The skinnier of the two glanced at the Jeep following. North wondered if word had spread among the islanders of his fight with Drummond on the quayside. A place this small, everyone would know everyone else's business.

'Did *you* volunteer?' His glance taking in all of them, and they each nodded.

Under the green roses, Joy's face wore a look of surly intensity. 'Each does as Enoch expects and through our sacrifice of finite time and sacred energy, we move closer to ascension and to the next simulation.' He wondered if Gaffney felt that way when he castrated himself. His eyes went to Joy's arms, covered over with old scars. She made no attempt to pull down her sleeves. 'Still, right here, right now, Drummond is a fucker.' Her voice was quieter now. 'Don't turn your back on him.' That, at least, made sense. He should have brought the gun with him and to hell with the risk of being caught with it. At least it would have given him a fighting chance.

The truck was slowing, the engine puttering to a halt way too close to the edge of the cliff for North's taste. Was Drummond planning to throw him over the edge? He glanced left to right, figuring out the landscape if he needed to make an exit. It was flat and empty, if you didn't count the rocks. As an escape route, it wasn't hopeful.

One of Drummond's crew – the one with the ginger sideburns North recognized from the tech hub – unbolted the tailgate, winking at North as he came forward. 'Mind how you go, pal,' he said, putting out his boot and laughing uproariously as North tripped, staggered and only just managed to stop himself from falling head first over the cliff.

So that was how it was going to be. Good to know, he thought.

But there was no hand to the small of his back, no shove out into the infinite. Instead, everyone ignored him, as they unpacked the Jeep and the truck. The guy with the ginger sideburns handed round bacon rolls and poured out mean servings of thermos coffee in plastic thimbles for Joy and her two companions. He ignored North.

Breakfast done with, Joy first and then the beanie men clipped themselves into their gear and started roping up. He watched the three of them chalk up their hands and Ginger took Joy's bag of chalk from her as if she wouldn't be needing it any more. North smelled Drummond before he saw him. A musky tang of expensive aftershave and machismo.

'Up for it?' Drummond asked, handing him a canvas bag attached to a belt.

North knew he had nothing to prove to Drummond. Drummond was a gangster, a bully and a sex trafficker, while North was a decorated soldier, a former assassin, and a government agent. He should refuse to play his assigned role in whatever dangerous game this was.

'Try stopping me,' he said. And there was a burst of unholy laughter from Drummond, who slapped him on the back harder than North had thought possible. The big man waved Ginger over and gestured to North. A climbing harness was thrown at him and, cursing, North stepped into it. Carabiners were clipped to him and ropes snaked this way and that. He clipped the belt around his waist and took a deep breath.

North watched as Joy eased her way over the edge of the cliff. Her face was a study in concentration. No fear. She was looking forward to what was going to happen. He peered over. She was standing on a ledge. It looked beyond perilous to him.

She worked her way down and the beanie-hatted men followed her, passing one to the other the poles with the rope nooses.

The stink of guano in his nose was intense.

'Over you go then, sunshine,' Drummond said, and pushed him off the edge.

North felt himself swing and then drop past the other climbers, down further than he'd thought possible, and smash side-on into the rock face. Pain shot up his arm. But he'd been lucky. The rope had caught on an overhang, which had absorbed most of its length. If Drummond had meant to kill him, he'd misjudged it.

And North wasn't giving him a second chance.

He looked up, past the shocked faces of the three climbers, onwards to Drummond's malevolent grin. The rope attached to his harness was feeding through an anchor point close to the top. It must have been in the rock already. His eyes travelled sideways along the line. The two men were clipped into another one further along, but Joy had free-climbed a good distance down.

What were they doing?

He attempted to swing himself closer to the rock. Before his fall, the other climbers had been intent on what looked to be some kind of hunt for seabirds. Now they saw he was all right, their attention switched back to their own situation and on the baby birds Joy was snagging and then passing back to her companions, who were stuffing them into wicker panniers and their own canvas bags. It was noisier down here than on the clifftop, he realized. The shrieks of the gannets bouncing off the rocks, the wind in his ears.

Squinting, he could make out Drummond's head. He had to be prostrate on the ground, with one of his heavies holding him by his belt. Sunlight glinted off the knife in his hand and

North thought of Gaffney. Drummond was straining to reach down to the anchor point. Tiny bits of gravel dusted his face and he turned it away. Drummond was working to ease the anchor point out of the rock.

North used his weight to attempt to move himself further away and then closer in to the rock.

Above him, the three other climbers moved horizontally across the rock face. They knew what was happening now, he realized, and they didn't want to be anywhere near him when it went down. When he went down. They wanted deniability. No, they hadn't seen a thing. Yes, it was a terrible accident. Please don't kill us too.

North's feet and fingertips grazed the rock and much as he wanted to reach for it, he needed to swing away again.

More gravel. He glanced up again. One of his heavies had handed Drummond some kind of tool. He was attempting to ease it behind the steel plate of the anchor point. The knife can't have worked.

North had to judge this right or it was over. Below him, the noise of the North Atlantic hurling itself against the rocks travelled upwards, loud enough to make his ears ring.

Above him, he imagined Drummond getting a purchase and wrenching the tool backwards and forwards.

He grabbed for a handhold on the rock and his toes found a ledge, tucking himself into the stone as the length of rope slammed into his head, the steel plate pinging and clattering against the rock.

A volley of laughter and expletives from above that North hadn't fallen – yet.

A distant shout from Drummond. 'You might as well let go, North. Don't fight the inevitable. Death comes to us all.'

There was a piercing whistle and the three climbers spread

out on the cliff were still for a second. None of them looked down to meet his eye. His boot lost its footing on the narrow ledge and he jammed his fist into a fissure in the stone, feeling the graze of it along his skin.

'Oi, you lot! Up the rest of you come.'

'We haven't got enough birds,' Joy called back up, casting North a worried look. 'This is the best spot. We need more time.'

Drummond sounded angry. 'There's a better place further on. Do as I say.'

The three other climbers were watching each other. The beanie men didn't know him. They weren't getting themselves into trouble for the sake of a stranger.

'Joy,' North called up to her.

She appeared to be thinking. She was one woman and Drummond had three of his crew with him. She was unmoored from the rock. All it would take was for them to start throwing pieces of rock down at her and it would be over for her as well as him.

He sensed as much as saw Joy pass the wooden pole up to the men and the men pass it up to the group at the top of the cliffs. They started climbing back up to the top. Joy the last to go.

He called out to her, but she ignored him. Or appeared to. Under her breath, she called down and over to him. 'Don't panic – that's important. Watch where I climb. There's a rhythm to it. And trust your instinct.'

His fingers were being wrenched from their sockets. His breath coming fast and hard. He felt as if he was never going to be able to move again. He tried to concentrate on where she was placing her hands and feet, and he could tell she was taking it as slowly as she could for his sake. There was a shout from the heavies at the top telling her to get a move on, and she called back up that she'd hurt her hand, but she was coming.

Drummond's face appeared, as if keen to get a last glance at his predicament.

'Hey, North,' he said. 'Kincaid sends his regards.'

After a brief delay, he heard first one engine and then another start up. They rolled off into the distance. There had always been a chance that Drummond would wait for him at the top. But Drummond felt he didn't have to. As far as he was concerned, Michael North wasn't coming back because any second now he was going to meet with an unfortunate accident while making himself useful.

Kincaid? He could understand the helicopter fly-by. But this? Kincaid hadn't just warned Drummond that North wasn't what he seemed. He'd told him to see he met with an accident. Not that Drummond needed any encouraging. It didn't really matter which one of them he should blame right now. He could deal with that later. If there was a 'later'.

Lovely. He swore under his breath and a gannet shrieked a final protest at the disappearing baby seabirds. Drummond, he decided, was a bastard of the first order.

'All right, let's do this,' he said, peering upwards. He had a choice. The fastest route was in a straight line from where he was to the top, but he had no idea of the lie of the land. Alternatively, he could traverse the rock face to where Joy had been snagging the birds and attempt to do what she'd advised – follow her route back up. It would mean going through the main colony of birds and also risked tiring himself out before he even started the vertical climb.

He felt paralysed. Either way, he could fall to his death. The icy-cold spray of the sea hitting the rocks was across his face and over the backs of his hands. Down here, the rocks were wet and slippery with it.

But indecision would kill him soonest.

The rock face didn't want to kill him, he told himself. The rock face was indifferent to his fate. It was Drummond who'd set him up to die and there was nothing new in a man wanting him dead. He found the thought reassuring. Frankly, he'd be doing his job wrong if someone didn't want him dead.

The weight of the dangling rope still attached to his harness was pulling him down. Inch by inch, he pulled it from his harness and let it fall onto the rocks below.

For a second, he considered what Linklater would say if she could see him right this second. Or Fang. Nothing repeatable.

He reached out a hand and started moving.

The gannets rose as one as he approached them, and the smell got worse. The rocks were slippery under his feet with their guano. He kept moving. Wiping his fingers on his chest, ignoring the pain in his arms and legs. He took a breath on a ledge, spreadeagling himself against the rock, his legs trembling.

Onwards, he told himself. Upwards.

It was difficult to decide which was the greater motivator – the promise he swore to drive his fist into Drummond's face at the first opportunity or the urgent desire he felt not to fall screaming to his death on the jagged rocks. In the immediate years after he had taken the bullet, he didn't care if he lived or if he died. It made living with the idea he could die at any second easier. But those days were gone. Now he had things to live for. People to live for. And every second was precious, because remarkable things could happen between one breath and the next. That was where hope lived, he thought. After you had taken a breath, there was always the hope you would take another. That was where life was – not just in the breath but in the hope.

Grunting, he swung himself upwards. Joy had started climbing around this point. She'd been fearless – making huge strides and balancing on what appeared to be shadows.

It's all in the mind, he told himself, staring up at what looked like a sheet of the sheerest rock he had ever seen.

40

London

It was barely seven in the morning when Fang rapped on the door of the Georgian house in Kennington. The house seemed like any other in the square, aside from the fact it had iron bars on the inside of the windows and two small palm trees stood on the step. They had been chained up, one to the drainpipe and the other to the cast-iron railing. Beyond the Georgian square, the hum of city living and diesel engines, and in a neighbouring house, a baby was crying.

They had left Granny Po knitting in the car and listening to the *Today* programme on Radio 4. She rather liked the limousine, Fang thought.

Plug stood next to her, his eyes swivelling constantly. He wasn't happy. Fang pushed the doorbell and, behind the wood, heard a penetrating trilling. She kept her finger on the buzzer.

'You and me out on the road again. You realize this makes you my sidekick,' she said to him. An attempt to lighten the mood, which was good of her, she figured.

She knocked again. Generally, she did not see making people feel good about themselves as part of her job description.

He peered down at her from his great height then took over on the doorbell. 'I'm nobody's sidekick, China girl.'

'I've told you before, that term manages to be both racist and patronizing. Plus, I was born in Newcastle.'

He raised an eyebrow. 'Fair enough. "Snowflake Fecktard Geordie" it is then.' He started whistling David Bowie's 'China Girl', and it was Fang's turn to scowl. Did he consider himself her bodyguard or her babysitter? Or both? Maybe that was one of the reasons Granny Po loved him? She sensed that urge to protect in the big oaf. When would these macho types realize that brute force didn't always win the day when you were up against enemies like Roxie?

Behind the door, she sensed movement, and felt Plug come to attention.

Since she'd told him about the CIA, Plug had moved on to a state of high alert. Judging by the pouchy bags under his eyes, he hadn't slept. He'd wanted to move her and Granny Po out of the country so they couldn't find her, but she already knew that wasn't the answer. This was the CIA they were talking about, and the most it would mean was a delay. And she didn't need delays, she needed answers.

'Who is it?' the voice called out from inside.

'Fangfang Yu and an idiot,' Fang said, and Plug harrumphed his displeasure. 'I emailed you. I'm sorry we're so early.'

There was the sound of lock after lock after lock being undone and the chain rattled on its track as a harried-looking Lucy Sampson opened the door and peered first at Plug and then down at Fang. 'You emailed me? I don't remember that,' she said. 'Let me check.'

She shut the door on them again. Fang pressed her ear to the wood and thought she caught the sound of rustling paper. 'Nope,' Lucy called. 'I can't find you in the system. If you're not in the system, you can't come in.'

'Could you please open the door, Dr Sampson,' Plug said.

'We can't make out what you're saying. And we brought bagels.'

The door opened a crack again, and he slammed his boot in the gap and placed his hand against it. A bag of bagels dangled from it.

'Please talk to us,' Fang said. 'I'm in trouble and I think you can help me.'

Fang couldn't see much, but had the sense the woman was shaking her head. 'If you're in trouble, I'm the last person you should come to.' She made to shut the door again. 'I can't remember what day it is.'

Fang looked up at Plug and nodded. As if it weighed nothing at all, the big man slowly pushed the door backwards – there was the sound of screws ripping from wood and a metallic chink as the security chain hit the stone-paved floor of the hallway. Lucy Sampson fled down the hallway, and the Post-it notes covering every inch of the walls rustled as she passed, one or two dropping to the floor. There had to be hundreds of them, in all different colours and sizes.

'I'll have to send someone round to fix that now,' Plug muttered as they stepped inside.

They followed Lucy down the hallway and into something that was once a kitchen, where an ancient computer stood on the kitchen table cum desk. Everything reeked of cat litter that needed changing.

Fang squinted. Lucy was writing the words 'Fang' and 'Idiot' on a Post-it, which she stuck on the kettle. She turned around and smiled as if she had forgotten they had forced their way in and presumed by their presence that they were friends. 'I have a memory like a...' She gestured with her hand.

'... sieve?' Plug said.

'... 'n elephant – if I write everything down. That's what

my brother used to say anyway.' She blinked several times. Fang didn't know if it was a nervous tic or because she was struggling not to cry.

It might have been early, but she was at least dressed. In her early fifties, she wore a pair of denim jeans, with oversized rainbow-coloured glasses on her small beaklike nose, while another pair held back a mop of shining silver curls. A third pair was tucked into her stripey top where her cleavage might have been if she'd had more flesh on her. A fourth pair dangled from a beaded necklace. And a fifth pair had been slid into a brooch in the shape of the moon.

Lucy noticed her looking and her hands went to each pair as if checking to make sure they were all there. Once she'd done with the glasses, she removed a cat from her keyboard and sat down at her desk. Her gaze fixed on Plug's battered face and huge shoulders, and the blinking stepped up to machine-gun fire. A nervous tic, then. He was making her anxious. But that was a natural reaction.

Plug grinned at Dr Sampson. 'I'm harmless,' he said. 'And on the inside? I'm beautiful. Here, have a cinnamon and raisin bagel.' He put one down in front of her, winked, and stepped away from them. He crossed to the window, the bag of bagels still in his hand. He pulled one out, licked off a dollop of escaping cream cheese, took a bite that halved it, and peered through the glass, checking out the back garden. Fang didn't like to tell him that he wouldn't see the CIA coming.

Fang waved a hand under Lucy's nose to break the hypnotic spell Plug cast. She could tell by the way he stood that he knew what she'd had to do and liked it. He told everyone who would listen that he had natural charisma. North described his friend as a people person. Fang held to the theory that people were only too willing to be charmed by Plug. Not least because they preferred the idea that he was charming to the possibility he was about to beat them to death and eat them.

Lucy looked at the bagel and then at Fang, as if puzzled by both.

'We're here so you can tell us about your brother Laurie and what he did on Murdo,' Fang said, keeping her voice even as if the conversation had been agreed and was only to be expected.

'Laurie?' Lucy repeated the name as if it was the first time she had ever heard it. Then she made a face. She scrabbled to pull up the sleeve of her top. Tattooed on to the inner skin of her forearm were three lines of italic text. *Benson killed my brother, Laurie. Read the file.*

Dr Sampson gestured Fang across to her desk. A bulging file with the name Laurie on the front cover. She tapped it. 'I put this together while I was able to. I wrote myself a note explaining never to put it away. Always to leave it out where I can see it and find it easily. I keep hard copies of everything in case they're across my system.'

She gestured to the computer then seemed to take a closer look at Fang. Fang got that a lot. It was normally followed by some comment or other about her age. Instead, Lucy put her head on one side.

'How much do you know about space?'

'Some,' Fang said. Leaving aside her conversation with Paulie, Fang had once studied under an astrophysicist with an interest in cutting down radio signals being emitted from Earth. The only teacher she'd ever had who had recognized her genius and seen her as anything more than an oddity. Fang still missed her.

Lucy blinked rapidly as if it helped her absorb information, but she appeared to take the teenager's word on it.

'I'm a space anthropologist at UCL. Or at least I was. I studied what we understand to be life, whether we are alone in the universe and what that means, how people live in outer space – which is surely coming. Laurie was an engineer. He designed satellites and he'd worked for all the big names in aerospace. He was my twin. When we were kids, we only spoke to each other up to the age of seven. It drove our parents bonkers.' She opened up a pendant at her throat. Like the brooch she wore, it was in the shape of a moon. A man's face smiled out of it. Like Lucy's, but shy and soft rather than confused. With a tiny click, Lucy closed it up again, her hand around it as if it were all that mattered in this world. 'Where was I?' Chewing her lower lip, she gazed at Fang. Plug had made them all a cup of tea. He placed a cup close to Lucy, close enough for her to notice but not so close she would knock it over. She gave him a nervous smile.

'You were telling us about Laurie and who killed him.'

A look of acute distress flitted across Lucy's face. 'Laurie's dead?' she said, appalled.

'This could take a while,' Plug said, settling into an armchair. 'Check your arm, love,' he said, taking a noisy slurp of tea.

Lucy checked her arm, gasped, and then stared down at the file. She opened it. Fang could see the Post-it stuck to the inside cover. It said 'Get help'. If anyone ever needed help, she figured Lucy Sampson did.

She gave Lucy time to flick through various pieces of paper. There was a handwritten letter she herself had signed that took her five minutes to read through. 'You need to take this,' she said finally. 'According to this letter, although I have flashes of clarity, I'm getting worse. The doctors say there's nothing to be done for me. Pretty soon, I won't care what happened to Laurie any more because I won't remember he ever existed.

But if what's here is true, I don't want them getting away with what they did to him.'

Fang nodded. She wondered if Lucy had thought about the fact that when they took the file, the tattoo on her arm would make no sense. There would be no file to help her remember. Just the catastrophic news that her brother was murdered. Fang extracted the contents from the file and slid it all into her rucksack. She took a piece of paper from the printer and wrote on it. 'You got help and gave us the file, Lucy. We'll deal with Benson. You don't have to worry about any of it any more. Laurie was your twin and he loved you.' She wondered how to sign it. And whether it mattered. In the end, she didn't.

Did Laurie Sampson love his sister? She had no idea. But if not, she figured it was a lie worth telling.

In the hearse, Fang ate her own smoked salmon bagel. The rucksack next to her on the backseat. Granny Po had moved into the front to be closer to Plug.

'Do you think the Benson people will come for Lucy?' he asked, his eyes finding hers in the rear-view mirror.

Fang chewed and swallowed. 'Judging by what I saw in the kitchen, she's been a regular visitor to their offices, and sent any number of letters and emails, so I'd guess it's possible. Maybe even probable. Today's Thursday. Gaffney went bonkers at 8 p.m. on Tuesday night. Lucy was back at Benson at midnight that night. If we're right about there being a connection between the Narrow Yett and Benson, I'm only surprised they didn't come for her yesterday. She's certainly made a lot of noise about her brother's death. They'll hate that. If they killed her brother for whatever reason, what's to stop them coming for Lucy?'

A beat.

'Then we have to get her someplace safe,' Plug said.

Fang took the last bite of her bagel as she stared at the passing traffic and the people going about their daily lives without a care. Because there were people like her and North and Plug looking out for them.

Above all things, Lucy hadn't wanted to forget Laurie. It had been a catastrophic wrench for her to give them the file in the certain knowledge she was giving away a piece of herself. Fang hoped her trust wasn't misplaced. She licked her finger tips to clear them of crumbs and thought about all the Post-it notes stuck to the hallway, which Plug hadn't read but she had. They said things like:

Your name is Lucy Sampson.

Remember to eat.

Remember to drink water.

Have you turned off the stove?

Have you got your key?

Your purse?

Check you are wearing your watch.

Lock the door.

Check door is locked.

Drink water.

Drink tea.

For bus, turn right outside door. Number 49 takes u to town.

And:

No file = kill yourself. Do not hesitate! It's time.

In the scraps of sanity she had left, Lucy had constructed a plan. Get help. Hand over the file. And when that was done – die. It seemed brutal but reasonable to Fang. A plan that she herself would adopt if ever faced with the same situation. Lucy had been a brilliant woman. Now she lived in utter confusion without the person she had loved best in the world. Dying seemed a rational response. There were worse things. She wondered if Lucy would manage it. She thought she would, especially if Benson came calling.

'I think she'll be okay one way or another,' she said, pulling the rucksack closer and reaching into it. 'Right now, we need to go see Bald Paulie at the foundation.' She drew out a bundle of papers and started reading.

'One way or another?' Plug said. He drew the car to a stop and turned in his seat. Granny Po turned too, and glared at her. 'What the hell does that mean? Start talking, China girl.'

41

Murdo

It was the smear of blood North saw first. He wasn't sure how she had done it, but somehow Joy had managed to cut herself badly enough to leave blots on the rock. She couldn't blot every step and handhold, but it was enough to guide him upwards. That had to be why they had taken her chalk bag, in case she marked the way. He had to give it to Drummond. The guy was organized. But first he had to move sideways along the rock so he could follow her route. He took it slowly, at one point using one hand to break an egg and drinking down the gelatinous gloop. He made himself swallow it. Refusing to retch. Did it again. He'd passed the night doing heavy-duty construction work and needed the boost. Protein was protein.

Risky reach after perilous foothold, his mind cleared itself of everything but the momentum of his own movements, his own breath, and the tension and stretch of his aching muscles. Till the pain seemed part of the climb and he became part of the cliff and the cliff became part of him. Time passed, the seabirds seemed to accept him, he moved with the rise and fall of the wind, till the moment he lost his footing and his hold on a nub of rock both at the same time and swung out over the gap, attached by one handhold, his eyes going down to the waves, and there was a sudden jolt of adrenaline.

What's the worst that could happen?, he allowed himself to think. It was still better than Linklater taking a saw to his skull.

You could fall, moron-person, he imagined in Fang's voice, and it was enough to bring himself back to the wall of rock.

I can do this, he told himself.

'I can do this,' he told the rock.

It was a long, slow climb up to prove it.

At the edge, he reached up – first one hand, then the other. First one forearm, then the other, his torso flat against the grassland, his legs still hanging off the edge of the rock face that had nearly claimed him.

For a second, he allowed his eyes to close and his cheek to press against the horizontal earth and horizontal felt like heaven.

Which is when the figure in front of him stepped onto first one hand, then the other, and pressed the soles of their polished boots down onto North's bruised and bleeding fingers.

'Hello, North,' Kincaid said, spitting out his toothpick. North saw his reflection in each lens of Kincaid's aviators. 'I figured you'd be harder to kill than Drummond thought you'd be.'

The muzzle of Kincaid's gun was pointing directly down at him, his index finger on the trigger. The gun looked comfortable in his hand. Was that a good thing or a bad thing? Kincaid wasn't an amateur so he wouldn't fire in error. No, Kincaid was a professional security operative and former soldier. He'd only ever fire when he meant to.

Kincaid's weight felt as if it was crushing the bones of his fingers to powder. As the toes of his own boots scrabbled for purchase, North knew there was no room for error. He would have one chance. He raised his right foot and found some purchase on a small shelf. Praying it would take his weight, he drove himself upwards and forwards and over the edge all in one thrust. The surprise of it made Kincaid stagger and as the American stepped backwards, North's arms reached around the other man's legs, his skull taking a glancing blow from the grip of Kincaid's gun as both men went to the ground with a resounding crash, the impact dislodging the aviators from Kincaid's face.

Sprawled over him, North's hand grasped Kincaid's right wrist. On no account could he let Kincaid raise that hand and fire the gun. Even if he didn't get the angle he needed to wound or kill North, the noise would be certain to bring other Americans to the clifftop.

Kincaid's rat-like teeth in the burnt face were bared. His hand reached up to push against North's jaw. North felt his neck creaking under the strain of it. He smashed his head full force into Kincaid's nose, blood everywhere. But Kincaid was made of strong stuff and his hand went back up, clawed and ready to take North's eye.

North pulled his head away. He wasn't getting it. Hone was proof enough it wasn't a good look.

Instead, he smashed Kincaid's wrist down on the ground, once, twice, three times, but the man still didn't let go of the gun. The American was smaller than him, but he was hard muscled and strong. More than that, he hadn't just spent a night doing construction work and he hadn't been dropped down a cliff face and had to complete a heart-thumping desperate climb back up to what he thought would be safety. Kincaid twisted, turned, then drove a knee into North's groin and used the surprise of it to roll out from under him. He brought the gun up again.

This time, he was breathing heavily. With the back of his forearm, he wiped away the blood that was streaming from the broken nose and across the shiny burn tissue.

'Is this over that incident in the bar?' North said, getting to his feet before Kincaid could think to keep him on the ground. 'Because this seems a tad excessive.' The helicopter fly-by and the flip-off he understood, but not this.

Kincaid hawked out a gobbet of blood. 'I'm over the bar. "Bygones" and all that.'

'In that case,' North said, 'care to tell me why you're about to shoot me.'

'Actually, I'm going to give you a choice. I can shoot you but that'll mean questions if your body's found. Or you can walk off that cliff, which won't mean any questions at all. For me anyway. You met with a horrible accident.'

'You'd rather I walked off the cliff?'

'I would.'

'I'll walk off the cliff if you tell me *why* I have to walk off the cliff. How about that for a deal?' He wasn't walking off any cliff. The guy was going to have to shoot him.

Kincaid's free hand dipped into his pocket and he brought out another toothpick. He slid it between his teeth. 'I can work with that.'

'You know who I am? What I am?'

Kincaid nodded.

'How?'

'My boss got told about Narrow Yett Jonny cutting people up on Tower Bridge. Got told your government was sending someone over. I figured it was you right from the get-go when you picked a fight in the bar that you could never win. I figure you did that for a reason – to get yourself a lift and a cover story as to why you needed it.'

'And you're prepared to risk the questions when my people come looking for me. I'm a government agent. They will come looking.'

'What do I know? You're here bothering the Narrow Yett nutters, aren't you? I met you in the bar, but I never saw you once you got here.'

'Mac will say you flew over us and instead of helping you left us to drown. That speaks to ill will.'

'He's an old man who went through something traumatic. I have plenty of men in that bird with me who will testify it didn't happen and we saw nothing. I couldn't believe my luck when I saw you in the sea. I thought you were going to save me a lot of trouble.'

'Nothing in what you've said so far explains why you want me dead. I'm not walking off the cliff yet.'

North could hear the wind at his back and the roar of the waiting sea. He'd survived a drop from Tower Bridge into the Thames. Somehow, he didn't think he'd survive this.

'I hear you were a soldier once. Ever serve under a great man?'

'I served with all kinds of men,' North said.

Kincaid's eyes were faraway. North knew the look. The memories of sand and comradeship. Of Life and Death. 'Teddy Benson is a great man. A great warrior. He's going to be President one day. No one wanted to know me or my guys when we came home. Teddy looked out for us. Teddy's guys don't have to wait for medical treatment. He makes sure we get what we need when we need it. He gets us jobs. He looks after us and we look after him. Teddy doesn't want you on this island.'

'He told you to kill me?'

'What Teddy wants, Teddy gets.'

North tried to remember what Fang had said about the Benson Corporation. That they ran the satellite operation here. That she believed there was something hinky about what they were doing.

Apparently, she got that right.

'Our cameras picked you up when you came to recce the base. You were mighty interested in one building in particular.'

And there we have it, thought North. Fang was right. As ever. The Americans had something shady going on in that building.

'Care to put me out of my misery? What's going on in there?'

'If I told you, I'd have to kill you. Oh yeah! I'm killing you anyway, aren't I.'

42

New York

It was three in the morning New York time and Mrs Benson was having a light supper before bed. Foie gras on four tiny pieces of melba toast and a single glass of Krug. She had her own geese on a farm in Connecticut. Had watched the pipes rammed down the birds' throats and the workers pour down the grain and fat so that their livers swelled to ten times their natural size. She felt a flicker of pity for the dumb creatures but not so much that she would stop eating the foie gras. She could, of course, do what most people do and buy it in, but Mrs Benson believed in quality in all things. And how best to achieve it – control the process from start to finish. She had employed the same approach to being a mother.

Teddy had been sound asleep when his mother's butler, Alan, called him and he knew better than to protest or be late. He was as familiar with his mother's erratic timekeeping as she was.

As he approached, she raised her face to his and kissed him on the lips, as was their habit.

'You look beautiful, Mother.' He was correct, of course, but the compliment carried little weight. Teddy blessed her with one of his most winning smiles. She knew that smile. He employed it when he had done something he knew she would disapprove of. Of course, she knew about the latest teenage escort. It really was too bad. Not an aspect of his personality she had planned for. How was she supposed to guess that the boy would have such primitive urges? Perhaps she should have left him in the army longer. But her timetable for his career progression was absolute. She was still considering how to deal with his appetite. These things did have a habit of creeping out these days. The only solution – she had almost decided – was the import of a steady stream of girls from foreign parts. All of whom would have to die. She had a part of the Connecticut estate already set aside for it. Actually, a rather nice spot. And once he had agreed, and she was sure he would agree, she would take the bodies in their plastic wrapping from the butcher's freezer where they were currently hanging and bury them too. She wouldn't discuss that detail with him. It might irritate. And Teddy was a bore when irritated.

But for now, she was more interested in this Murdo business.

Without being summoned, Alan brought Teddy a cup of black coffee, a rare fillet steak and scrambled eggs mixed with cream and chives. Teddy didn't thank Alan, merely smeared Dijon mustard on the side of his plate with a yawn. She would have to listen to him chew, which she hated, but that couldn't be helped.

She waited till he had carved himself a large chunk of steak and had his mouth full.

'Teddy, darling boy. Why do you keep things from me? It hurts my feelings.' She raised a sparkling finger to the corner of her eye as if stopping up a tear. They both knew there was no tear.

Teddy swallowed, then gave a sharp cough as if he had swallowed too soon and the piece had been too big. He took a mouthful of piping hot coffee, which had to burn his mouth. Was he wondering if she was talking about the escort? As if she'd care. She cared only because his predilections introduced an element of risk to the very long game she was playing. One she had privately dubbed How To Grow A President.

'What do you mean, Mother?' Teddy said, his look one of charming bafflement.

He was an excellent liar, she thought, which was fortunate, bearing in mind he would have to lie through his teeth to win an election.

But she knew him backwards and forwards and sensed him stiffen in anticipation of her challenge.

'I'm a doddery old lady, I know, darling.' He murmured his denials, as well he might. She was remarkable for her age. Remarkable for any age. 'Did you think I wouldn't find out?'

A flush was rising under that fine skin. She had to stop herself from laughing out loud. She would hardly confront him about paedophilia and serial murder in front of the help. Help did love to gossip.

'The hack, dear boy. What are you doing about the hack? You realize we cannot have hackers romping through our systems, accessing commercially confidential information, technical drawings, financials. The next you know they'll be issuing ransom demands and passing our latest missile systems to the Chinese.'

He gave a sigh of relief. 'The hacker has been identified and shut down, Mother. It's some kid.'

'Isn't it always,' she said. 'Don't these children read books any more? Go for walks?'

'Not this one,' he said. 'She works with the British security service.'

Mrs Benson picked up her glass and took a sip. She left a crimson mark behind. Mrs Benson didn't like complications.

'Are you telling me the British government is behind this hack?' Her voice was flinty.

'No, Mother, I've been assured that isn't the case. The kid did a trawl because of the cult guy on Tower Bridge. She knew we had the spaceport out there and was poking around. The agent she's working with on the island happened to have a run-in with some of our security guys up there. We've done everyone a favour. Our people are pumped to have snagged her. Apparently, the kid is brilliant.'

Mrs Benson began to tap her crimson nails on the damask tablecloth. 'I don't like mess, Teddy.'

Teddy sat back from the table. He knew how this went. He closed his knife and fork and Alan's hand materialized to take his plate away.

'Mother, it's being dealt with. The hacker is ours now. The British know not to complain when she disappears. And the guy on the island is about to meet with an unfortunate accident.'

'And the engineer's sister?'

'Mouse is handling it.'

Mrs Benson knew that already. But it didn't do to let Teddy know how often she and Mouse talked. 'Then I want that young hacker on the plane over here today.'

43

Murdo

Kincaid opened his mouth. Closed it. Turned his head. North had watched Mia's arrival on the quad bike, but the American had been oblivious. As North talked, out of the corner of his eye he'd watched Mia stand to get a closer look at what was going on. Watched as she switched off the engine and start pushing it towards them. He'd watched the bike gain momentum and then grabbed for Kincaid's aviators, scooping them, swivelling them and trying to plunge them into the meat of Kincaid's thigh. They shattered in his hand.

'What the—'

Kincaid's hand went to his leg as North dived to the side, hitting the earth, and rolling far and fast.

Distracted, the American was not so lucky. The impact of machine against the man – his scream hanging in the air for what felt like minutes after his body had disappeared over the edge, spreadeagled over the front of the quad bike.

North moved into a sitting position on the hard ground, stretching out his arms and flexing the fingers of his cramped hands. His nails were torn and ragged. Grazes covered the skin and his knuckles were red raw. 'I was glad to see you before. I want to state for the record – that feeling does not go away.'

Mia reached down her hand and pulled him up. Pain from

his scratched and aching fingers. The face with its familiar brown and white patterning, the dark eyes and ringlets blowing in the wind. She'd saved his life. For the second time. The first time, she'd been twelve and he'd been thirteen and she'd saved it by showing him there were decent people in this world, and if someone like Mia could be his friend then he wasn't a worthless piece of scum.

'What is that dopey look on your face, North?' she said. 'Because we do not have time for any...' – she waved her index finger in between them – '... silly business.'

He laughed. She had that effect on him, he realized. She made him happy. Mia had every excuse to do nothing with her life after the start she'd had. Instead, she was a creature of light. He was glad her adoptive parents had loved her. She deserved to be loved. He was glad she was on his side. Mia was the most loyal person he knew. She wouldn't be on this island of crazies if she wasn't. And she was brave. She'd been here five months without any support – emotional or practical – and without an ally other than Rowan.

North screwed up his face as if to say he had no idea what she was talking about. 'I was wondering if you're okay. You did just run that guy off the cliff.'

She gasped. 'Oh my Lord. You're right. But he was going to kill you. Am I going to get into trouble? Maybe he isn't dead?'

She moved to the edge of the cliff and North moved with her. They peered down. The body lay splayed – limbs higgledy-piggledy – on a jagged rock below, eyes fixed and staring, the broken-apart head to one side, waves breaking over and over it, rising up and smashing down, blood and brain matter mixing with the sea foam and spume. In a couple of hours the ocean would take it all.

'What was his name?' she said.

'Kincaid.'

North stooped to the ground and picked up a wooden toothpick and tossed it over. The guy had really liked his toothpicks – it seemed a small enough thing to do for him. Glass crunched under his boot and when North moved his foot, a lens from the aviators reflected back the memory of Kincaid's face in a hundred splinters. He moved his foot back and pressed down some more, treading the memories into the earth.

Mia put her hand over her mouth. North reached out and patted her back. 'He wasn't a nice person.'

She raised her huge dark eyes to him – a glimmer of a smile. It was what he used to say when he hit someone in the children's home. He checked his watch. It was 8 a.m. – he had four hours till the broadcast. The sooner this mission was done with, the sooner he could get off this island. He had enough on with the cult, he didn't need the Americans pitching in. He'd been lucky Mia showed up. 'How come you're out here?' North said.

'I figured Drummond would pull something. I came to see if you needed help. I think he's planning to leave. Word is his men have wrecked the computers in the offices pulling out the hard drives, and they've been shredding like crazy. No one dare say anything because his people are armed and deeply unpleasant. But everyone down there is getting freaked out. Something is off, and I can't get any sense out of anyone.'

North nodded. 'Kincaid told Drummond who I am and who I work for. So it makes sense that Drummond's shutting up shop. He must have figured out it's over for him here. The place is going to be crawling with police soon enough. I'll deal with him later, but I have to get back to the cottage, grab my phone and warn Fang.'

'Hold on one minute.' She gestured to the edge of the cliff. 'I know his name now, but why did he want to kill you?'

'Because someone called Teddy Benson ordered him to.'

Mia's look was one of puzzlement. 'The guy who runs Benson Corporation? But why? You're here because of Jonny Gaffney and the cult.'

'Benson is the parent company of D'Urberville, who own the spaceport. Kincaid worked there. After I had that encounter in the bar, I asked Fang to look into them. They have a building here that isn't on any plans. I recced the spaceport and took a look-see at that building when I got here. The cameras picked me up, and it turns out they are hypersensitive. They have something going on there – I have no idea what. But if they're willing to kill me, then I'm guessing Fang has either found something or she's about to. I have to warn her. Once I've done that, I'm taking out the tech hub. I hope for Enoch's sake he really does believe this isn't reality. He might care less when I kick him clean out of it.'

44

London

Mouse went in with two of his best people at nine. He'd
planned to go in earlier, but he had held off to talk to Kincaid.
Held off and held off. Because there was no sign of Kincaid
anywhere on the base. Mouse had a bad feeling.

Lucy's cats started mewing and Mouse wrinkled his nose at
the stench. Plus, he was allergic to cats. This time, he'd taken
an antihistamine but even so, his throat started to itch as soon
as he made it through the back door. Two cats started winding
themselves around his legs and, holding his breath, he shook
them off. He hadn't been impressed with himself yesterday. If
he had arrived just a little earlier, if he had completed his search
quicker, if he had simply managed to lift the files before the
carer arrived, he wouldn't have had to take the risk of coming
back. But it was too late for regrets. He was here now. He'd
brought help and he wasn't leaving till the job was done. Then
he could go back to the problem of finding Kincaid. At least
the CIA woman, Roxie, had come good with the key. He had
had his tech people check it. The key was clean. The kid had
acquired a lot of information in a short while, but nothing that
couldn't be handled. Especially once Lucy was taken care of.

Not that he was a big admirer of Roxie. The woman was
a toady and a psycho – he could read it in her. And what was

it with that hand? And those tiny teeth? Not like Mrs Benson. Now she was a woman to admire. Brilliant. Wrinkled and old as she was, Mouse found her wildly sexually attractive, and Mrs Benson sensed it. Not that she would ever – not that he would ever. Still, the idea of their union entertained him during his otherwise dutiful sexual encounters with his wife. A nice woman called Sandy who had given him two children, a daughter currently at an Ivy League school and a fifteen-year-old boy who played quarterback whom scouts were already sniffing around. Mrs Benson always remembered to ask after their welfare. Teddy? Teddy didn't even know he was married.

He made a gesture to Brad and Brett for them to go ahead of him, blinking back the water in his swelling eyes. The air felt like half oxygen, half fur. As they moved, the walls rustled with a thousand Post-it notes. *Keys*, he glimpsed. *Feed cats*. He'd lock the cats outside of the house when he left, he decided. He wouldn't want them feeding on Lucy. Lord, but losing your memory was a terrible thing. It had happened to his grandfather years ago. The guy would go walkabout in his pyjamas. Drove his grandma into an early grave, worrying about him. Left the old guy behind when it was really him who should have died. He hadn't been kidding, Teddy, when he said they'd be doing this broad a favour.

And in a way, Mouse admired her. Broken brain or not, she knew something was up with her brother's death and she wasn't letting it go. Good for her. Of course, loyalty like that was the reason he was here. Loyalty like that meant Lucy had to die but maybe the brother would be waiting for her on the other side. He didn't think so. But it was a nice thought.

Brad placed his hand against the door and pushed it open. They stood at the threshold.

There was the sound of a gentle rhythmic snuffle coming

from the bed, where a tortoiseshell cat also lay sleeping. The cat opened a green eye, shivered and jumped down from the bed, eased its way through their legs and made for the open door and the stairway. How many cats did a woman need? Mouse wrapped his hand around the meds he had in his pocket. He was hopeful he could use Lucy's own medications for the job. These were a top-up in case there weren't enough. Not that it would look like suicide. More a forgetful lady got altogether confused with her pills. He didn't have time for drowning in the bath any more. He looked towards the bedside table. No glass of water. Under his breath he swore, made a gesture to his two men to wait right there, and went back down the staircase into the kitchen for a glass. Holding his breath to avoid sucking down yet more cat hair, he turned on the tap and ran the water till it was cold.

45

Murdo

It took another hour to make it back to the settlement. The lights were on in the refectory, but North and Mia skirted around it. She pointed down to the quayside. Drummond's men were stacking bags and boxes and covering them with tarps. The crime boss was getting ready to pull out. And it looked as if he was planning to use the Zodiacs. North didn't envy him. It had been bad enough coming over on Mac's boat. It was a long, rough crossing via the inflatables. But with Drummond's men busy on the quayside, it gave him a run at disabling the tech hub. Drummond might even have given him a head start if he'd been stripping out hard drives in the offices.

Mia tugged at his arm. 'I need my gear. My camera and laptop at the very least.'

North nodded. 'Okay, but be careful. Once you have what you need, meet me at Rowan's cave. It's the safest place till Hone sends in reinforcements.'

He also wanted her a long way from the settlement. Because once he had done what he had to do in the tech hub, he was moving on to Enoch. And if that unwound, he wanted Mia and Rowan as far as they could get from Enoch's groupies. He watched her go.

His own cottage was empty. He unzipped the compartment in the rucksack and pulled out the satellite phone and the Sig Sauer, which he tucked into his waistband. It was a shame Kincaid had taken his gun with him into the sea. Another weapon would come in useful. He picked up the satphone and connected with Fang.

'*Moron helpline,*' she answered, her voice in a whisper.

'Why are you whispering?' North said, keeping his voice low.

'*I'm in a cupboard. Plug's here too. Well, he's in the wardrobe. How may I help you today? Can I suggest you try turning it off and then on again.*'

Cupboard? Wardrobe? He didn't ask.

'Fang, you need to watch yourself. The DUT security guy – a man called Kincaid – just tried to throw me off a cliff. Well, to be fair, he offered to shoot me if I preferred.'

He could sense he had Fang's full attention. 'Why?'

'He said Teddy Benson told him to. They served together in the army. He was something of a fanboy. Look, since I got to the island, I've been focused on the Narrow Yett – all I've done with the Benson people is scope out the spaceport, but he mentioned your mystery building. I asked you to look at them, so you need to watch your back. These people don't mess around.'

'*Too late, mate,*' she said.

'What?'

'*I said, it's getting late, mate. Where are you on the broadcast? You realize you only have three hours left till Enoch goes full crazy horses.*'

'I don't know where the main server is.'

'*It's probably a high-spec laptop they keep someplace central. It would be good to have, but you don't need it. Just*'

*kill the power into the place. I have to go. There's a moose
loose aboot the hoose.'*

'What?' It was a terrible line.

'I'll tell you later.'

'Fang, tell Hone I'm doing this and then he has to send over
reinforcements urgently. I need to get Mia and Rowan off this
island. Once I take down Enoch, I don't want these crazies
strapping us into a wicker man and setting light to it, right?'

'On it.' She hung up.

The meter box was at the rear of the building. He'd noticed
it when he'd met Mia in the graveyard. And he'd remembered
it correctly. He smashed the key with a rock and levered open
the door, wrenched the fuse and its carrier from the box amid
a fierce shower of sparks and closed the box up again. It was a
blatant act of sabotage and not one he'd be able to talk his way
out of if challenged. But they were past that point. Drummond
had tried to kill him – he was done playing nice with these
people. His first instinct was to throw the fuse and carrier as
far as he could, but there would always be the risk they'd be
found. Instead, he buried them in the loose earth of one of
the recently dug graves, patting down the turf afterwards. He
stood back. No one would notice the intrusion. Judging by
the numbers still on the quayside, the church wasn't highly
staffed. Anyone in there would be groping their way to the
doors right now. Even as he thought it, three young techs
stumbled out of the fire door. One of them made for the meter
box.

'What the hell!' he said. 'Did Drummond do this?'

That would be useful. Drummond had been stripping out
hard drives. It would make sense he might go bigger.

North was already climbing the metal-runged ladder up to the satellite dish. At the top, he swung himself over the ledge and onto the roof. Cables snaked this way and that over the roofing material. The satellite dish was huge, bolted down and protected against the ferocious winds. He didn't see there was any way around it and it was going to make a lot of noise. He got out his gun and started firing. Six shots into the centre of the dish. Judging by the sparks, it was enough. Loading a fresh magazine, he could hear shouts from the men below. They were cult members, not Drummond's men. If they came up by the ladder, he didn't want to hurt them. He made for the door into the building, wrenched it open and plunged into the darkness.

Fang thought the server would be a laptop. If she was right, then he had a good idea where it was – the main atrium. Ahead of him, he glimpsed a phone being used as a torch. Whoever was using the phone was making their way along the corridor towards the old church and the exit, but it had given North an idea. The phones that had been taken from the Narrow Yett members on their arrival were lining the wall of the edit suite where Drummond had watched Mia's rough-cut footage.

At the thought of Mia, he wondered if he had done the right thing letting her go off like that. Should he have kept her with him? But she was surely safer out of the settlement, especially when Drummond believed she was running a con. The last thing North wanted was Drummond deciding Mia needed dealing with.

North found the door. But it took three attempts to find a phone with enough battery to give him a decent light. Time was running out. Any second now the place could fill with people attempting a fix on the electricity supply. The main atrium was empty. The laptop wasn't obvious. He opened every cupboard in the place till he found it. Fang would be impressed if he brought it back in one piece to gut for intelligence. Even as he considered it, he heard the noise. He was turning when he felt the blow to his head. Then there was nothing.

A second later and he is everything and everyone and everywhere. A younger North, the familiar weight of a machine gun cradled in his arm, the feel of compressed sand under his boots, and air that baked your lungs.

And then it happens. Death comes for him. Helpless, he watches as a bead of sweat rolls from the forehead of the Afghan sniper, sees himself in the sniper's sights, feels the squeeze of the trigger against the calloused finger. He wills the finger to ease away but he can't stop history.

The spring mechanism hammers against the firing pin, which hits the primer, and it's like it's hitting him.

The gunpowder explodes and he feels it in every nerve.

He is the swell of gas in the cartridge case and just like that, the bullet pulls away and shoots along the barrel of the gun and he does too – the bullet explodes into the air and flies. He does too. Time slows.

He sees himself again, watching the rooftops and the windows. The bullet moving ever closer.

Duck, he wants to shout. *Move. Watch out.*

But he can't.

And infinitesimally slowly, the spinning bullet fights its way through the wind and gravitational pull of the earth to meet the metal of his helmet, parting his hair and pushing through the bone of his skull. He is in his own brain now, watching synapses and neural networks spark and part as the bullet carves its way through his grey matter, before finally coming to a halt. He stands in his own brain as it moulds and mends itself around him. Electricity firing off everywhere. Pain.

Panic. Fear. The desire to survive. Before he is pulled back through the channel left by the bullet and he sees the brain close up behind him and he is flung out into the universe. Stars everywhere.

46

London

The inside of Mouse's nose felt red raw and he was fighting the urge to sneeze. That would be ironic, he thought. Blowing the operation because of a sneeze. He expected his men to wait, but he could hear the tread of their boots across the bedroom floor. In his gloved hand, the surface of the water in the glass trembled. Mouse was irritated. He didn't want Lucy panicking at the sight of strange men in her bedroom. He didn't know if she would remember his face. Probably not, but at least he knew how to talk to her. He was pretty convinced he'd be able to coax her into taking the medication. And if not, she would take it anyway.

The men had at least held off from waking Lucy. Brad stood next to the right side of the bed, Brett at its foot. It was a double and Lucy was curled up in the centre underneath a flowery duvet covered in pastel pictures of what looked like baby deer and rabbits. Treading softly, Mouse approached the left side, put the glass of water down on the bedside table, and clicked on the bedside lamp. Nothing. The bulb must have blown.

Then again, he didn't need much light for what was coming next.

'Lucy,' he said, quietly but loud enough for her to hear.

'Lucy!' He stroked the shape of her arm under the duvet and waited for her to stir. 'Lucy!' Louder again, as the figure in the bed sat bolt upright, a gun in her hand pointed straight at him.

Everything happened at once. The doors of the wardrobe flung themselves open and an ogre appeared. Roaring, the ogre reached first for Brett, whose back was to him, and then for Brad, who had started to turn. The ogre propelled one man into the other and there was the sharp crack of skulls smashing into each other. Both bodies dropped to the floor.

Mouse's attention switched back to the old woman in the bed. She appeared unperturbed by the ruckus. She continued to chunter and squawk at him. Mouse had no Cantonese. But he had a pretty good grasp of what the old lady was saying and it wasn't *Good morning, looks like another lovely day.*

There was a small cough. A young girl stood in the doorway, her hair in stubby plaits, Joe 90 glasses, glittery boots and a quizzical look on her face. Mouse knew exactly who she was. Fangfang Yu. And knew too that Mrs Benson had been right. The teenage hacker hadn't stopped poking around – despite Roxie Powell's threats.

Fang shushed her grandmother, although the old lady continued to mutter. The mutterings going up a notch as the ogre gently extracted the gun from her small liver-spotted hands. Mouse wasn't sure who he preferred to be in charge of the gun. The old woman or the ogre. They both looked like using it wouldn't be a problem.

'You've been filmed breaking and entering this house, Mr Munson.'

How did she know his name? Had she searched company records? Did she realize he had served with Teddy? Above all things, he had to keep Teddy out of this mess.

'And by the way, you know your shoe squeaks, right?' She

flashed a grin revealing turquoise braces. 'Next time you sneak up on someone, maybe wear trainers.'

It wasn't his shoe. It was a metal rod in his ankle, but the kid didn't need to know that.

'You're still on camera now, in fact. The feed from every camera installed in this house is being recorded by a friend of mine, whom you'll never find. If anything happens to Lucy Sampson – ever – my friend will release the footage to the authorities, the press and broadcasting media. I understand Benson's reach is wide, therefore my friend will also release it to all social media outlets in this country and in yours. Your identity and your position within Benson Corporation are known. I don't know the identity of your two friends here, but I will soon. I'll use the footage to run a facial recognition programme. I imagine they've served in the US army. I don't think it'll be hard to discover their identities – social media is a wonderful thing. I'm afraid that Lucy no longer possesses what you came here looking for. Her condition is worsening every day. Soon, she'll have no memory of her brother or her brother's death. I'm sure you agree with me that she should be left to live out her life in peace. However long that life is.'

Mouse nodded. What choice did he have? None. His nose was beginning to stream again. He turned his head, raising his arm a little to wipe his nose. And the granny in the bed gave a snort of disapproval.

As it happens, he was humiliated but not entirely unhappy at the way things had turned out. After all, Lucy was only doing her duty by her dumb-ass brother. But this kid had no idea what awaited her. He contemplated saying nothing, but he didn't want there to be any doubt in Fang's mind.

'You were told to leave this alone, kid. I'm going to have to tell them what went down here and they're not going to be happy with you. You're in a whole lot of trouble.' She was the same age as his youngest son. And just as his own son would have done, she rolled her eyes.

'What's new?' she said.

47

Murdo

When North surfaced back to consciousness, it was dark and something was wrong. Very wrong. It took him a second to remember where he was and where he should have been. He was in his cottage on Murdo. He had wrecked the power into the tech hub and the satellite dish. He'd been trying to grab the server. And he should have been en route for the manse and Enoch.

Leaning over the side of the bed, his head screaming with pain, he retched, vomiting up bile, which splattered onto the stonework. His eyes felt heavy in his head, but he forced them open and kept them that way till they got used to the flickering darkness. What time was it? His eyes went to his watch. A minute or two past eleven. If he had it right, he'd been out cold for around an hour and a half.

He'd stopped the broadcast – at least there was that.

But Enoch was still alive.

Someone had lit a peat fire in the grate and it still burned. Gradually, he came back to himself. What was this? And why had someone knocked him senseless and then brought him back to the cottage? And even as he wondered it, he felt a slithering up the length of his right leg.

North froze. He barely breathed. He had been stripped naked and something was in the bed with him.

More importantly – where was Mia? She'd said she was getting her gear from her cottage. Was she safe in Rowan's cave or had she been intercepted? North felt a wave of panic sweep over him.

He'd taken a blow to the head. Linklater had said he needed to rest up and look after himself. He'd ignored her. Before he came round, he'd been convinced he was fighting in the desert. Been sure he was dying. Was he hallucinating? His brain swollen and inflamed? Was this the start of the end?

Shadowy flames licked the plaster walls of the cottage as something slithered across his stomach and into the warmth of his side. Something else took hold of his left ankle like a hand encircling it. This felt all wrong.

If he pulled back the blanket and there was nothing there, it would be bad. It would mean he couldn't trust himself. That Fang had been right and he should have stayed in hospital.

The smell of the peat fire was warm and vegetative in the enclosed cottage. Was he imagining things?

A slow crawl started up in the valley between his ribs and his left arm before mounting the bicep and inching up his arm towards his face. Straining his neck, he tilted his head a fraction of an inch to stare down into the dark black eyes of a snake. His breath caught in his throat.

He was lying in a bed of snakes.

That was good, right? He wasn't imaging the sensation, which meant that his brain wasn't about to kill him.

Or was it? Did the snakes exist?

Because if you could hallucinate the sensation of snakes, you could certainly hallucinate the sight of snakes.

But this didn't feel like a hallucination. His body was warm on the right, the side close to the fire, and cold on the left, the side close to the wall. Surely, no one would hallucinate a detail like that.

He used his tongue to lick his lips and swallowed hard.

The snakes really were real.

And that wasn't good. Not at all. He didn't like snakes.

He thought back to the Narrow Yett symbol of the serpent curled around the gate. To Mac's talk of those who left, broken – 'Tales of snakes and beatings and goings-on I don't even want to understand'.

They had taken his gun and they had taken his satphone. But if they'd found him doing damage in the tech hub, why not just kill him? And this didn't feel like Drummond. This was too twisted. He was meant to suffer. Otherwise, why leave him in a bed of snakes? Were they poisonous? Would they bite him? That had to be a risk if he moved too fast. They were comfortable, lulled and sleepy courtesy of the warmth of his body. They'd only woken because he had woken and moved. Now they were curious. Wary. Which made them dangerous. He had to keep still and think. Otherwise, his next move could be his last move. Whatever he did, he had to make sure he didn't startle them, because if he startled them all bets were off.

Slowly, he moved his head. Was there any chance his gun was anywhere? He'd had it to shoot out the satellite dish. But there was no sign of it.

There was a gust of freezing cold air as the door to the cottage was flung open and someone entered. Behind the new arrival, the wooden door screeched along the stone floor and then closed with a bang.

48

London

Laurie Sampson was a bright guy. As his sister said, he'd worked as head of engineering for any number of big satellite companies, some in the US, some in Israel, one in the Far East and one in the UK. Fang kept reading, then pulled out her own laptop.

She checked her Snapchat. Lucy was en route for a home within a newly created dementia village on the south coast. It was the best available, with 24/7 care, its own shops, spa, a restaurant, café, lake and woodlands. She was allowed to keep two of her cats and Fang had already set up a trust in Lucy's name to pay the bills for the next thirty years. Lucy would have the proceeds of the sale of her Kennington house. She'd also inherited Laurie's estate, and in the event the money ran out, Fang would cover the rest. Benson Corp would, of course, be able to find her if they looked; Fang believed they wouldn't look. In a way, she felt bad Lucy hadn't been allowed to follow her own plan. But when she'd told Plug about the Post-it notes announcing Lucy's intention to kill herself, he had point-blank refused to let it happen. Chuntering '... *unbelievable...*' and '... *not happening...*' and '... *God, give me patience...*', he'd swung the limo around and sweet-talked Lucy into a black cab driven by a friend of his, who'd driven her home to his wife.

Paulie had found the nursing home and made the calls and Plug's wife had insisted on driving Lucy down there herself. Aside from the two cats, among the personal effects they'd packed up for her was a photograph of her and Laurie when they were kids. Fang wondered how the nursing home would explain away the tattoo.

With Lucy dealt with, they'd been able to lay their ambush for the Benson people. They didn't have long to wait. Fang grinned at the memory of the face of Teddy Benson's henchman as Granny Po sat up in bed cursing him out in Cantonese.

But that was then and this was now.

Fang was set up in Paulie's office at the Derkind Foundation, close to St James's Park Tube station. She sat at a long table she guessed was usually meant for meetings, her glittering boots up on the chair next to her, while Paulie worked at his desk looking round and bald and bothered, the light from his desk lamp bouncing off his head. Behind him, his shelves heaved with robots, Star Wars figurines and white owls. A Harry Potter-style movie poster hung to one side, with Daniel Radcliffe's face swapped out for Paulie's. The poster read: Bald Paulie and the Future of the Human Race.

'That's so cool,' Fang had said as she walked into the office, and Paulie had beamed. She'd tried to imagine North beaming if she attempted a wind-up like that. Unlikely.

It turned out that Laurie Sampson did love his sister. He sent her regular texts, which Lucy had printed out, and he sent her letters. The texts had nothing very much. They referred to phone calls that doubtless Lucy would have no memory of. Asked questions about what the doctor had said about Lucy's memory. Reminders to eat and when he was going to call her. The letters, however, were warm and intimate, full of childhood memories and gossip about colleagues and chat about the weather. Tidbits about space exploration. He seemed to miss Lucy a great deal and he was obviously concerned about her. The letters, too, included cuttings of funny stories from the papers he read, bad poetry and song lyrics, puzzles and crosswords of his own devising – apparently a habit that had carried on for decades. As if he used them to exercise a different part of his brain. Which intrigued Fang.

This is the crossword I'm proudest of, Lala... His pet name for his sister. *I think my work here is done. I doubt I'll write another. In any case, I'm so busy I won't have the time.*

He was right about that, she thought. Within a week, he was dead in a car crash on a stretch of remote highway in the Highlands. His car a fireball that could be seen for miles. It was all in the cutting from the local paper, which someone must have sent Lucy. Fang was about to discard the cross-word puzzle when something about it caught her eye. It was blank. She wondered why Lucy hadn't filled out the cross-word. Perhaps she'd been too upset? Or perhaps the disinte-gration of her mind had made it impossible. A clue caught her eye.

26 down. In the sky with Diamonds.

She sucked on the tail end of her plait. The answer had to be *Lucy*. As in The Beatles' classic, 'Lucy in the Sky with Diamonds'. It was too easy. She should put away the puzzle and figure out what she was going to do about Roxie, then start over to make sure she hadn't missed anything. But before she did that, maybe, since she was here, she should ask Paulie what he thought she should do about Roxie? But then, if she did that, it would turn into a whole thing. L U C Y she filled out the answer.

Surely the clues couldn't all be that easy? Laurie had said he was proud of it. As if it was a thing of intellectual beauty and rigour.

Fang knew the theory behind building a crossword puzzle, although she had never bothered herself. But she knew you could have a theme to them and if you did you could often tell what it was by the longest answer. Was Laurie proud of the theme he'd selected for the puzzle perhaps?

Her eyes scanned the puzzle for the longest answer. There were four ten-letter answers. Two of the clues gave her the words 'suspicious' and 'assimilate'. Was Laurie perhaps instructing his sister to pay close attention to the answers in the puzzle? Fang sat up straight in her seat.

8 across. Sweet Mickey Rourke movie.

Sweet? As in candy? As in sugar? As in sixteen? The sixteenth movie? Or one he made in 2016? She wasn't too proud to search for the answer. *The Wrestler* didn't fit. The one he made in 2016 fitted perfectly.

Weaponized.

The black-and-white squares of the puzzle seemed to shift and reshape themselves. Letters flitting in and out of squares. Settling then taking off again. Chasing each other up and down the lines.

20 across. Stormy little seas.

Fang let her mind wander. Crosswords sometimes relied on anagrams.

Stormy was the kind of word that implied being tossed around. What if she moved the letters of *little seas* around?

She sat up straighter in her seat.

Satellites.

What did she know so far?

Lucy believed Laurie was murdered.

It was true to say that Lucy's faculties were compromised by her memory issues. But that didn't mean she was wrong.

An attacker broke into Lucy's home and tried to kill her. It was her memory that was at fault, not her logic.

And if Laurie was murdered, it was for a reason.

Did he know he was in danger? Because if he did, it made sense that he would try to pass a message on to his sister.

Fang went back to the puzzle. Working her way through clue by clue.

23 down. Extraordinary sense of being. (Lat)

It was some Latin word or term.

In esse.

It meant in existence.

It took her fifteen minutes. Which was a good ten minutes longer than she expected.

She scanned the crossword, pulling out the words that seemed to have some relevance. Generally speaking, Laurie seemed to have used the 'across' clues to pass on his message rather than the 'down' clues.

Lucy. Space. In esse. Weaponized. Satellites. Lasers. Solid. State. Comms.

She narrowed her eyes and, if she was right, the word 'almost' was apparent in boxes that were the palest of grey instead of white.

In her head the words advanced and retreated, constantly moving. She let out a small gasp.

Lucy, Space (is) Weaponized. Solid State Lasers Almost In Existence (on) Comms (communications) Satellites.

'Look at this.' She held up the crossword puzzle. Bald Paulie came out from behind his desk and crossed to the table. She moved her feet and he sat next to her with a whoomph. He took the paper.

'Well, that's not good,' he said.

'Tell me what you know about solid-state lasers and satellites, Paulie.'

Paulie screwed up his face, presumably clearing down what he had been working on to make room for her question. He reached forward to his table of reports, picked one up and rifled through it. He pointed out an appendix.

'See here. The US Navy have demonstrated a solid-state laser known as the MK2 MOD 0. It's supposed to be the most powerful laser out there, with an output of 150 kilowatts.' He glanced back up at her. 'That's a big step up from the system they used to have, which had an output of around thirty kilowatts. It would be enough to take out drones, small boats, et cetera. The advantage of a laser weapons system is you don't need to go get more ammunition, all you need is the fuel to power it. A solid-state laser is powered by some kind of battery system rather than fuel and that makes it cheaper again, more efficient and durable.'

'Are you sure we don't have them in space already?'

'Absolutely not,' Paulie said. 'They aren't operational down here yet. Nor are they effective enough to take out larger aircraft or missiles.'

'What if I told you they were about to have them in space? Why would they be sent up there?'

Paulie sighed. 'We've had five decades' research and development into lasers. What you're saying would be premised on a massively higher kilowatt output, and if that was the case...' She waited him out. It was as if he didn't like to think about where his logic was taking him. 'Then the most logical target would be "enemy" satellites. And they wouldn't just be "blinding" them, they'd be destroying them.'

Paulie hesitated again. He stood with some effort and started pacing. 'Space warfare is the next frontier. It's incredibly easy to envisage a war down here spreading into space,' he said. Fang knew his work into artificial intelligence had led Paulie down all kinds of dark alleys into the consequences of tech for humanity. She wondered how he was still able to spread kindness and joy when he spent so much time anticipating global catastrophe. Or maybe because he anticipated catastrophe, he felt obliged to be kind and joyful in whatever time humanity had left.

'Everybody is as interested in it as they are in AI. Putting weapons of mass destruction up there is illegal. But, as you and I already talked about, satellites are already used for geo-observation, recon and communications support. They're used as part of any nuclear command system. Modern warfare involves space. Looking ahead, if you can knock out enemy satellites, you have an immediate advantage in any combat scenario on Earth. Courtesy of history and the sheer scale of their investment, the US still maintains pole position up there. But Russia, China, and the usual suspects are all going to keep pushing to even up the playing field.

'The Americans haven't been happy for a while about the way Chinese and Russian satellites manoeuvre too close to their own key satellites. Sometimes that's about intelligence gathering and sometimes it's about forcing the American satellite to move on. It could be – but I'd hate to think that this could be the case...'

'It could be what, Paulie?'

'If the US beat everyone else to solid-state laser technology in space, they would have the option of a pre-emptive strike – something that could knock out everyone else's military satellites. Or more than that – they could take out every satellite that even had the potential to provide comms or mapping. I'm talking commercial GPS, everyday mobile comms, satellite TV. You name it.'

'They couldn't do that. Russia and China wouldn't stand for it.'

Paulie frowned. 'No, they wouldn't. But that doesn't mean to say that there won't be hawks pushing for it stateside. It's bad.'

Fang's eyes went back to the note Dr Laurie Sampson had scribbled on top of the puzzle. At first reading, it looked like nothing at all. Now it looked like a message from beyond the grave.

This is the crossword I am proudest of, Lala. Because it was the one with most significance. *I think my work here is done.* He didn't mean he'd finished constructing the puzzle. He meant he'd finished with D'Urberville Technologies. *I doubt I'll write another.* And he knew they were coming for him. *In any case I'm so busy, I won't have the time.* Because time for him had run out.

She imagined Laurie Sampson out on the highway. Already dead in his car. Someone like Mouse or Roxie making sure it looked like an accident before they set light to the car. The explosion into the night sky. Was Laurie trying to make a run for it when they caught up with him? Had he made them suspicious? He must have been desperate as he constructed his puzzle, trusting his sister and her fragile mind to make sense of it.

When she looked up again, Paulie's back was to her. He had opened a window and was looking up into the sky. Space was not supposed to be weaponized. Was he asking himself exactly how many death rays were on their way up there?

49

Murdo

The boy stood pale-faced and dripping rain on to the stone floor.

'Stop right there,' North ordered.

Rowan's voice was tremulous. 'The storm's coming off the sea and the cave's running with water. Can I stay here for a while? I won't be a bother.' He took a step into the room, and another, and North barked, 'Rowan! Stop!'

The teenager started retreating. Step by step, as if he feared North might well rise from the bed but only to beat him.

'You have to get out of here, kid.' North kept his voice even and quiet, but even so he sensed the thrum of it was waking the snakes who hadn't already woken. 'The bed is full of snakes. I'm not joking, and I don't know how many of them are loose in the cottage. They could be dangerous.'

'Snakes?' Rowan let out a small gasp. 'You see, this is what I told you about. You must have done something to hack Enoch off. This is one of his ordeals. For followers who let him down.'

Rowan took a huge breath. If it was possible, he was even more nervous than North. But instead of turning tail and heading straight back out the door, he stayed exactly where he was and appeared to be looking around the cottage, as if searching for something. He found it. He pointed at the wicker basket that had earlier been full of peat but which now stood empty on the hearth. 'We can put them in that.'

North had to admire the boy's courage. But then the significance of his presence hit him like a punch to the throat.

'Rowan, where's Mia? Isn't she with you?'

The boy seemed to get even paler. 'I haven't seen her. Is she all right?'

North didn't reply. The hope he'd had that Mia was out of the settlement and safe from Drummond and from the madness that was the Narrow Yett was gone. But if she wasn't with Rowan, where was she?

A light sweat had broken out over North's body.

'Rowan, you can leave. I don't want you in here. It's dangerous.'

Rowan shrugged. He appeared to brace himself. 'Not if you handle them right.'

North swore under his breath. He didn't want Rowan putting himself at risk. But he couldn't get to Mia until he got himself out from under the snakes.

Rowan was crossing the floor, step by careful step. 'If you displease Enoch or those closest to him, especially if you aren't honest with them, they dump the snakes on you. Enoch uses them in his services sometimes too. Then the faithful get to handle them. That's considered an honour. This – not so much.'

A dry slithering started up close to North's ear and slid over his neck to curl up on the pillow close by his head. A tiny wet kiss from the snake's flickering tongue touched his cheek.

'Some of them are de-fanged.'

Hope leaped in North's chest then died back immediately.

'"Some" of them?' He spoke without moving his lips more than he had to.

'And they milk some of the others before the services.'

'"Milk?"' North felt he was repeating the boy like an idiot, but he figured these were unusual circumstances and his brain was taking its time to process the facts.

'Of poison,' Rowan said. He reached out to the chair where North's clothes lay in a tangle, before wrapping North's hoodie around one arm and his jeans around the other.

Step by careful step, he approached the bed.

'I don't have a stick,' he said as he lifted the tiny snake by North's cheek and carried it over to the basket by the hearth. 'I should really have a stick.' North heard the snake hit the bottom of the basket with a small thud and a hiss.

Scanning the floor, the boy carried the basket over to North's bed. His lower teeth sucked down his top lip and he lifted the sheet and blanket that covered North's recumbent form. 'Wow!' he said, his eyes widening.

'That good, huh?' North said.

Moving as little as he could, North raised his head enough to wish he hadn't. Snakes of all shapes and sizes were draped and curled over his body.

'They really seem to like you,' Rowan said, keeping his voice low. And North found himself impressed that the boy could manage humour at such a moment.

'What if I stand up really fast?' North said, the desire to be out of the bed and out of the cottage and off the island growing with every passing second. Rowan lifted up a snake that was curled over North's penis, which was, North thought, a good choice, because being bitten by a snake anywhere was a bad thing, but being bitten by a snake there would be beyond unfortunate.

Rowan shook his head as he peeled away what looked like a baby python, which writhed in apparent protest before he dropped it into the basket.

'It's risky,' the boy said. 'One of the faithful got nervous

around that one...' – he pointed at a skinny green one –
'... and when it bit him, he had to go to the mainland. The
doctors cut off his arm at the shoulder.' He made a chopping
motion high up his arm.

North raised his eyes to the ceiling.

This was his fault for not arriving on the island, breaking
down the door of the manse and shooting Enoch dead where
he stood. He'd played it too safe and this was his punishment.

He sensed the weight of the green snake lift and as it did
it struck the boy's denim-wrapped arm with sudden and
unexpected violence.

Rowan dropped it into the basket and whistled. 'That is
one bad-tempered snake.' The boy appeared to have almost
forgotten his own fear, so focused was he on the job in hand.
North worried that it might make him careless with his own
safety, but the young face was a picture of concentration.

'This is bad, but it could be worse. Elvis isn't here.'

North tried not to think about each and every snake the boy
was lifting from him. The weight of it lifting. The shift of air.
The disquieted slither of the snakes left behind.

'Elvis?'

It was almost as if they were making conversation to distract
themselves from the horror, North thought.

'The rattler.'

Nope. Here came the horror again.

'Elvis has already killed two men and one woman. They're
buried on the island.'

Were they the graves he'd seen in the graveyard with the
stones piled up, North wondered.

'But we'd know if Elvis was in here, because he rattles like
stink.'

Thankful for small mercies, wasn't that how the saying

went? North breathed a sigh of relief and as he did, a distinct and ominous rattling noise started up from underneath the bed.

50

London

Hone preferred not to work out of the MI5 building in Millbank. Instead, he had his own suite of offices that operated out of a Tudor timber-framed building in Holborn. Although there was modern art on the walls, there were no personal photographs or anything that passed for memorabilia, with the exception of a tuning fork on a small wooden plinth. Sitting across from him, the idea came to Fang that Hone had used this fork to kill the man who had taken out his eye. She had no proof of this. Of course, that didn't mean she was wrong.

He nodded towards the tablet she had perched on the corner of his desk, which flashed with images of people and snakes and rockets. 'What's going on there?'

'I've designed a program to analyse the images on Gaffney's wall to see if they tell us anything.'

'Why?'

She shrugged. 'Why not?'

'I'm more interested in what you're telling me about the fact the Americans are about to put up weaponized lasers into space.'

Fang moved the wad of bubblegum she was chewing to the side of her mouth and caught it between her lower teeth and her cheek. 'Laurie Sampson worked for a company that is about to put weaponized satellites up in space. That company is a subsidiary of a major supplier to the Pentagon. That means the US government is about to have weapons in space, and under the cloak of commercial providers launching their own constellations of satellites, the Americans are indeed on the brink of populating space with laser weaponry. The kind of laser weaponry that isn't supposed to exist. .

'I think it's fair to say that if other countries with satellite interests find out what the US has been doing, they're going to start doing the exact same thing. Not least because they will know it's possible. They'll catch up and they'll do it fast. The dangers of this escalating are huge.'

'And how do you know this?' Hone said.

'Laurie's sister Lucy gave us a file on her brother. In it there's a crossword puzzle that he sent her. The last puzzle he ever sent her. Laurie worked for D'Urberville Technologies as one of their most senior satellite engineers. We believe he was involved in integrating solid-state laser weaponry into what everyone thinks are communications satellites. Satellites that are about to be sent into space. When he lost his nerve and tried to quit, they killed him for it.

'D'Urbeville Tech, or rather its parent company, Benson Corporation, isn't freelancing. I think any work they're doing on this is with the full knowledge and cooperation of the US Space Force or the Pentagon. Certainly, the Technology and Innovation Office of the Space Force signed off on their spaceport.

'Murdo is remote enough to seem insignificant in their operations. Bearing in mind Murdo's location, its satellites are only going into low Earth orbit and on a polar trajectory, so they couldn't do that much damage. They'd be travelling north–south, or south–north. Maybe they could take out imaging satellites or a constellation of communications satellites? But the big boy stuff – that's higher up in geostationary orbit. The military and governmental command and control of nuclear weapons, for instance. And that makes me think the technology will get rolled out across who knows how many launch pads over the US and into all kinds of commercial satellite operations. The US will have weapons all over space.

'I'm adding two and two here, but bearing in mind the UK is keen to boost our own defence capabilities in space, along with the speed of the planning permissions and licences granted to the Benson Corp, I'm guessing the British government knows exactly what's going on up there and is on board with it. Maybe we're getting some of our own in return for our cooperation? Maybe some of them will be operated in concert with the US if the time ever comes to use them?'

Hone got out a cigarette and Fang thought he did that sometimes to distract whoever he was with while he thought things through. Hone thought deeply but not as fast as she did, she realized, but then, aside from Bald Paulie, she didn't know anyone who thought as fast as her.

'Which means you already knew all of that when you sent North up there.' Fang shifted her weight on the uncomfortable chair and brought up her legs so she could sit cross-legged. She'd recently bought a soft black cropped vegan leather jacket. She slipped it off and hung it on the back of the chair.

Hone watched her get comfortable with a 'take-your-time-why-don't-you' expression on his face. 'That's all very interesting, Fang. But you were supposed to be looking into the Narrow Yett with North. I didn't ask you to look at Benson.'

Hone's gaze drilled into her. It was intense. She wondered if that was because he had just the one eye. Maybe he'd been an altogether different person when he had two. Maybe he'd been way more laid-back and honest? Maybe losing the eye made him shady? But she didn't think so.

'Shoot me, I can multi-task. I've bottomed the cult's finances. A gangster out there called Drummond has been washing dirty money. I've tracked back the bitcoin he's been buying to an IP number on the island. It's the final proof we were looking for that Drummond had his own reasons for being on that island.'

Fang would give Hone cred for listening with his full attention, which was a shame in a way because she wasn't going to be able to broad-brush anything. 'The thing is, we kept turning up links that made Benson Corp worth looking at. Firstly, they share the island with the cult. That's interesting in itself. Then, when I was trawling for what the company was up to, the Americans came knocking and told me to stop. Also, one of Benson's security people – a guy called Kincaid – tried to throw North off a cliff.'

'North can be deeply irritating. I understand the temptation. I can't imagine it ended well for Kincaid.'

Fang hadn't asked what happened to Kincaid. She'd assumed. She found it uncomfortable to be reminded how willing North was to kill people. She'd challenged him over it before. He said he had a rigid rule that if someone tried to kill him, he felt no compunction in killing them right back. 'If you make bad decisions,' he'd said, trying to sound virtuous, 'you have to live with the consequences.' This from someone who specialized in making bad decisions.

'But I'm just going to take you back to what you said about the Americans coming knocking.' Hone let out a plume of tobacco smoke and Fang felt her heart sink into her glittering boots. 'Who what why when where? As they say.'

This, she thought, was like all those occasions she had been summoned to talk to headteachers in the days she still went to school. Wrongly accused of disruption and disrespect. Of trash-talking teachers and showing up the fact they knew less than she did. Made to feel small and odd and worthless. Except this time, Hone might have a point, bearing in mind she had technically committed treason. This was bad, she thought. And it was going to get worse.

'Who? Roxie Powell, CIA. A right cow.

'What? She wanted me to keep her up to speed on what we found out about the Narrow Yett.

'Why? I messed up and they caught me.

'When? Yesterday.

'Where? Upstairs on a bus.'

'To my surprise, I find I still have questions,' Hone said.

Hone listened to her account of events, of Roxie's threats, and of her decision to pass on what they had found out about the Narrow Yett.

'Tell me why I don't throw you into prison right this second,' he said.

Fang's hands were balled up in fists where he couldn't see them. On no account would she let him see how nervous she was. If he didn't help her, she would be at Roxie's mercy.

'I didn't commit treason. Everything we had at that point, the Americans would have known already.'

Hone ground out his cigarette in a glass ashtray. '"Never work with children and animals." You being the child and North being the animal. What was I thinking?'

'There's a spaceport on Murdo. You knew I would have to look at Benson. You knew I'd hack their system and anyone linked to them. You also knew the fact they told me to stop looking would only make me look harder. I don't like you. But you're good at your job. If the British government is part and parcel of that particular space programme, then you already knew what I'd find in terms of the laser weaponry. But they're doing more on Murdo than sending up dodgy satellites.'

'Oh, yes?' His face was set, his lips a narrow line. Was he pressing a button under his desk for security to throw her in prison? She had to hope not. He'd do it himself, she thought. Pick her up and carry her down to some dungeon, throw her in, and slam the door on her.

'Because there's a mysterious building on Murdo no one knows anything about.' She used her phone to bring up the satellite image of the island on his computer screen.

He glanced at it. 'That proves nothing, and please don't mess with my computer. This is supposed to be a secure server. You shouldn't even be able to do that.'

'How about if I do this?' she said, typing furiously.

Hone's computer screen changed to a screen with a Benson Corporation logo. 'What am I looking at? And how many laws am I breaking right this second?'

'The memory key I gave Roxie gives me a backdoor into both the CIA and now Benson. A real backdoor. Not access to the fake pages the Americans put up in their sting. And the answer to your other question is – a lot.'

'Do I have to remind you the Americans are our allies.' Hone leaned over and pressed a button on his intercom. 'Jamie, get security up here. Young lady, you are in a great deal of trouble.'

Fang's eyes were on her tablet. 'Maybe, but I'm not the only one,' she said.

51

Murdo

North was willing to take his chances with the other snakes, but a killer snake with three corpses to his credit and a rattle like stink was a step too far.

North's boots had been under the clothes now wrapped around Rowan's arms. His instinct was to leap from the bed, grab the boots and the boy, hit the door and tumble into the driving rain. But Rowan was across the room by the basket. North took it slowly. Easing himself down the bed inch by inch to climb out at the end rather than the side. Even so, he stared down into the darkness under the bed, waiting for the snake to strike.

Rowan's eyes were fixed to where the noise was coming from. Sitting, North reached out a hand for the boots, picked them up and lifted them across. He pushed one foot and then the other into them and stood. The rattle had stopped. He took a step towards the boy. Then another. Three steps. Aware he was moving further away from the door.

'Rowan,' he said, reaching out a hand to the boy's arm.

Rowan didn't seem to hear him. There was the slither of the snake against the stone floor and the rattle started again. Glancing around, he saw the snake's eyes gleam in the darkness under the bed.

'Rowan.' He spoke softly. 'Veeeeeery slowly, I want you to move towards me.'

Another slither and the rattle. North risked another look. The rattler was out from under the bed and curled up on itself, the triangular head raised, the rattle shaking back and forth like something that had its own mind.

'We're going to die,' Rowan said.

'We're not going to die, Rowan.'

Were they going to die? He didn't know. He had to hope not.

'Kid, very slowly, step towards me.' They were in front of the hearth now. Nothing between them and the snake. Behind the snake, the bed. And Gaffney's mural with its bloody handprints and writhing figures and way too many snakes for comfort.

North had to drag his gaze from the painted snakes back to the real one.

Rowan took a step, two steps, three, till he was close. North moved Rowan behind him, so he was between the boy and the snake.

'Remember, take it slow,' he said. 'Do you think Elvis likes music?' He could feel the boy vibrating with nerves. The last thing he needed was for Rowan to make a run for it. He started humming, eyes on the snake, the snake's eyes on him. 'Love me tender, love me sweet, never let me go...'

The snake seemed to like it.

'You have made my life complete. And I love you so...'

They were at the door. North reached for the handle and pushed down. Just as the snake threw a loop of his body out and started to move.

North pushed Rowan out of the doorway, and followed him

into the street, slamming the door behind him. There was a thud of snake hitting wood.

North breathed out.

He was never listening to Elvis Presley again. It would bring back too many memories.

The boy had landed flat on his back and was winded. North pulled him to his feet and unwound his jeans and hoodie from the boy's arms. He pulled on first his jeans and then his hoodie, before kneeling to double knot the laces of his boots. When he stood, he took the boy by the shoulders, and stared down into the pale face. North gave the lad's shoulder a soft punch.

'Rowan, mate, you are impressively cool under fire,' North said, and the boy grinned from ear to ear.

'You don't much like snakes, huh?' Rowan said. 'Me neither.'

'In a wildlife documentary, sure I like snakes. In my bed? Under my bed? Not so much. Who do you think dumped them there?'

Rowan looked thoughtful. 'They're kept in the barns in special glass enclosures, it's all temperature controlled, and they have their own lighting and habitats, you know. After their keeper died, a girl called Comfort took over.'

Comfort had told him she used to work in a zoo. She didn't mention any reptile house. He'd refused to sleep with her. The snakes were what? Her revenge? He had misread her badly. She'd known who he was when she climbed into his bed. When she had gone up to ask Enoch whether he could stay, Enoch must have realized the arrival of a stranger on the island so soon after Gaffney's atrocity was no coincidence. Drummond was in contact with Kincaid. That would have confirmed it. Drummond must have told Enoch that Benson's man on the island knew for a fact North was a government agent. Enoch had told Comfort to get close to him and she had done her damnedest.

He was, as Fang always said, a moron. Distracted by a lovely face and a sweet smile, he'd even felt guilty about not being straight with her. The signs had been there all along. She was cold and vain enough to play games with him in the bar. She'd lain naked in his bed and said men always thought she was dumb. She'd cloaked her fury when he turned her down. But he should have read it. At supper she had milked him for information to take back to Enoch. Mia had called it right. She'd said the girl was hard and manipulative, and was a member of Enoch's inner circle. Comfort had lied from start to finish. Lied about wanting to take Gaffney to a clinic. She didn't care about Gaffney. She had made no attempt to defend or even feed Rowan. When someone shows you who they are, believe them the first time. Except he hadn't. And whatever Enoch was up to, Comfort knew all about it.

He started walking, and Rowan ran to catch him up. 'Where are we going?' the boy said. His teeth were chattering now, and North realized his original motivation to seek shelter had been forgotten in the drama. The lad was perishingly cold and soaked through to the bone.

'Do you know which cottage Mia is in?'

Rowan nodded.

'She's in a pod. They're smaller but warmer. She's the last pod in the row closest to the dining hall. Follow me.'

North didn't knock. He didn't have to. The door to Mia's cottage hung off its hinges.

The cottage had been turned upside down. Mia was bound

head to toe – pieces of her camera and laptop lay scattered across the floor.

'Mia!' He tore off the silver tape she'd been silenced with.

'North!' He sensed her trying to raise herself as he sawed at her bindings.

'Take it slowly,' he ordered her. He rubbed at her hands and her feet, trying to restore her circulation.

She winced, moving his hands away. 'I'm all right, North, really.'

'What happened?' he asked, but he already knew. Drummond had no intention of any footage of him or his men making it off the island.

She put a hand up to her head as if it hurt. 'Drummond was wrecking the place when I got back. He wanted my footage, so he slapped me about.'

Five months she had spent on the island. Five months' work.

'I'm sorry, Mia.'

He was going to kill Drummond for hurting her. For frightening her. For wasting five months of her life.

She levered herself into a standing position. 'He got the storage cards. But he's panicking about something. He wasn't thinking straight. Everything's already in the cloud.' She allowed herself a grin.

For the first time, she noticed Rowan and he went to her for a hug. She looked pleased, as if Rowan had never done such a thing. 'I'm glad you're all right,' he said. 'I thought you might be dead.' She patted his back. 'I'm fine. He wasn't trying to kill me. He said he didn't have to get his hands dirty.'

'What does that mean?' North said.

'I have no idea.' She had dragged a chair over to the wall and was standing on it, reaching on tiptoe up into the beams. 'Drummond didn't find this either.'

It was a satphone. Not identical to North's, but close enough. 'No way was I living on this island with no way to call for help. These people are insane.'

North took the phone from her. 'I love how you think, Mia Barton.' He pumped in Fang's number. She picked up after barely one ring.

'Thank God, I've been calling you on repeat. You need to get Mia and Rowan and get out of there. The police are on their way from the mainland. The navy too. Something about Gaffney's painting bothered me from the start but I couldn't put my finger on it. I wrote a symbology program...' North was barely listening. Where was Drummond now? Six hundred miles away and she sensed it. '... North, are you listening?'

He had to focus. 'Sorry, Fang. Say again.'

'Gaffney's painting on the wall of your cottage. Do you remember the symbol he painted in between the snakes?'

North pictured it in his mind's eye. 'Yep, the cult's logo. The gate inside the snake.'

'Really, a "yett" is made up of interwoven horizontals and verticals. Their logo isn't made like that. My program took it apart. Lifted pieces of it off and on again. Swapped them around. If you lift off the snake, what you are left with looks like a straight line, an oblique line and a right-angled triangle.'

He knew she was trying to explain something important, but he didn't have time for it. He had to find Drummond and rip his limbs off.

'You can't tell on most of their logos. You can tell with Gaffney's painting. Because he painted over that straight line twice. It's the letter K with the N over it. The snake around the K and the N forms a C.'

That's what Rowan had said in the cave. A snake like a 'C'. 'So what?'

'Enoch has planned this from the very start. KCN is potassium cyanide. Ingested, it can lead to nausea, vomiting, coma, seizure, cardiac arrest, fluid in the lungs, respiratory arrest and whole-body toxicity. You've stopped him making his call to arms. But he still has three hundred potential victims on that island. I think he's planning to kill them all.'

Down the line he could hear Fang's voice, the words not clear, but the tone urgent in what he figured was another conversation she was having in tandem with Hone.

'Hone says get yourself and Mia and Rowan to a place of safety. He says help should be there soon. Promise me you'll do that.'

What had Comfort said? The cult were celebrating the tenth anniversary of their arrival today. But what if Enoch was planning a party to end all parties?

As he hung up, North looked at Rowan. He could see fear on the kid's face. He had to find his baby sister and take her someplace safe. As for North, he had to save three hundred lunatics from the man they worshipped.

He pulled open the door and stared out at the street. The settlement was empty but from a distance he could make out a few people heading for the refectory hall, moving fast as if they were late and didn't want to miss out.

52

Outside Hone's office, the agents who worked as part of the Friends of Cyclops were looking at each other. They knew who the kid was and, judging by the noise coming from Hone's office, she had messed up big time.

He had already buzzed Jamie once. And three minutes later, two panting security guards had arrived. They currently stood on guard outside Hone's office.

Jamie's switchboard was heating up as North's assistant put Hone in contact with no one knew who.

'What do you think?' one said in a low voice to another. The other made a face as if to say he had no idea but was just grateful it wasn't him in there with the boss going full throttle at him.

'He's already put a call in to the Fleet Commander and now he wants to know how long till the army gets there. And he's reached out to the Americans for cooperation. That's all I can hear.'

Minutes after, Hone flung open his door. 'Please escort Ms Yu from this building. Now!' His voice carried across the room, although the security guards were right by the door. 'Or I won't be responsible for my actions.'

It was remarked on later that no one in the room had ever heard Hone sound so angry.

326

'Hone, let me at least help get North off the island.' The guards' hands had hold of her either side. She was the size of a sixpence. Complete overkill, the watchers thought as the kid's sparkling boots dangled down, the toes not even touching the ground. It wasn't as if she was going to give them any trouble.

'He doesn't need your help. We are done here. And needless to say, Fang, you're fired!' He slammed the door.

The agents nearest the windows watched as she walked across the square and under the archway out onto Holborn in her glittering boots. The kid never looked back.

53

Murdo

Rowan pointed at the harbour. Drummond's men were crawling all over the flotilla of Zodiacs. Loading them with bags and supplies. Drummond was shipping out and not losing any time about it. Rowan started to move towards them, but North gripped him by the arm. No one wanted a confrontation with Drummond more than he did. But they had bigger problems. And it wasn't as if Drummond had Tara down there. Drummond had his men with him but not his daughter and not the mother of the girl. It looked as if he was leaving both mother and daughter here to take their chances.

'I told you he didn't care about Tara,' Rowan said. 'If he loved her, he'd be taking her with him.'

Drummond, thought North, was caught between a rock and a hard place. He either left the girl here and trusted the mother to keep her safe, whatever was going on, or he took the girl with him on a perilous journey out to sea. A journey, it would seem, he felt he had to make. North kept watching as one of Drummond's men pointed up at the dark grey clouds and then out to sea where the waves chopped and churned. From a distance he heard the crack of the gun and watched as the dark-clad figure toppled face first into the sea.

Next to him, Mia gasped.

Drummond's other thugs gazed down at the floating body as if hypnotized, then, as one, turned their backs on it and went back to readying the inflatables. The Narrow Yett was breaking apart, he thought, watching them. Drummond was a bastard, but he ran the settlement efficiently if brutally – the tech, the money, the supplies, even the building they stood in right this moment. Now Drummond couldn't wait to get away.

If any of the community members had heard the shot, they preferred to ignore it. Drummond and his men had never been believers, and the believers knew it. Their faces were beaming, their smiles broad, and North could hear the anticipation in their murmuring conversation as they moved towards the refectory. The tramp of their boots on the cobbled road. Drummond had a gun and two dozen heavies with him. Could he still get Drummond to help him stop whatever this was? He didn't think so. Drummond was getting out and as far away as humanly possible from whatever this was. Because he knew it wasn't good.

He turned to Rowan, but the boy had already taken off for the refectory at a run.

'Rowan,' North yelled, but the boy ignored him.

He was supposed to get Mia and Rowan out of this, whatever this was. Mia was unzipping a side pocket of her trousers to bring out the satphone. 'It has no camera, but I can pretend I'm filming with it,' she explained. 'They're used to that. Follow me.'

North caught a glimpse of Rowan's auburn hair as the boy disappeared into the refectory. The boy was intent on one thing – Tara safe from harm. And he'd seen the chance he needed to snatch his sister.

The community hall was packed tight and the atmosphere charged. Some people sat crushed together and hunkered over huge plates of dark meat and bread. Others moved from table to table. And as each new arrival came through the door they were hailed, and as they found a place, they hugged and kissed their friends. Each group talking excitedly to the next. Husbands and wives holding each other. Men's arms around men. Women clinging to one another. Rocking together on the benches. Some laughing and some weeping. A heightened buzz and chatter of conversation and the sound of cutlery against china. The smell was rank and gamey – North guessed from the seabirds he'd helped to harvest.

Most didn't notice him or Mia. They were intent on their food or their companions. Those who did looked uneasy, perhaps because they knew about the ordeal by snakes? Or perhaps because Drummond had told them he was dead? He was a stranger come among them and they weren't mourning his passing, that was for sure. He'd have thought they'd at least have sent out a party to recover his body, but they hadn't even done that. They were all too occupied with whatever this was.

Across the hall, he caught sight of Joy and nodded – her hand bandaged from the gash she had given herself to help him. She looked down into her food as if the nod had been meant for someone else, but he thought there'd been a glimmer of a smile on her lips. He wanted to get to her but as he made a move towards her, the crowd eddied and swirled and North realized that the children were being led in a crocodile through a huge door out of the main refectory, hands reaching out to lay on the children's heads as they passed, as if in blessing. From where he stood, Rowan had seen the movement too.

As had Mia, who started pushing her way towards the door and the disappearing children. Whatever she planned to do for anyone else, she didn't want Rowan anywhere near this. But he was too fast for her. He too disappeared through the huge door.

Mia turned back. Her eyes desperate. She wanted to follow Rowan but she didn't want to leave North out here on his own.

His instinct was to go to her. If this was what he believed it was, he didn't want her out of his sight. But Mia's priority had to be Rowan. And he had to trust her to look after the kid and look after herself. He knew that anyone standing between Mia and Rowan wasn't going to be standing for long.

He gave her a thumbs up sign. Focus on Rowan. He'd handle everything else.

A look of relief passed over her face and she went back to pushing her way through the seething crowd.

They were standing and clearing the tables and the benches now. Pushing them to one side. He looked for Joy again but couldn't see her. It was like a rock concert, he thought. Or a political rally. The buzz of the crowd anticipating the arrival of the main act. The reason they were all there. And a feeling close to panic threatened to choke him. Surely these people knew better than to trust Enoch? But they were here on this island because they did exactly that. They ate when he told them to and worked when he told them to and hurt themselves when he told them to.

This was up to him now. He needed to access the cyanide and destroy it. He started moving people out of his way. Pushing and shoving them to one side, ignoring their complaints and protests.

But it was too late.

There was the sound of cymbals clashing and a rising crescendo of drums and fiddles as Enoch entered, and with him a fierce gust of cold wind. Even in a substantial sheepskin and leather coat, he was slimmer than North had been expecting. Inoffensive, with thinning hair and a smile that spoke of infinite patience. The crowd began to applaud and cheer as Enoch moved to the dais in front of the picture window.

Instinctively, North's hand felt for the gun at his waist, but there was no gun. Enoch's men had taken it when they'd knocked him senseless in the tech hub.

'My children...' A spotlight picked out Enoch. He'd taken off the coat and stood with arms folded in a grey jumper and black denim jeans. He was miked up, the rich mellifluous voice bouncing around the hall. North checked the beams; they were rigged with what looked like stage lighting. Checked the doorways. Each was guarded, the guards carrying makeshift bats, sharp knives tucked into waistbands, or held down by their sides. No one was getting out of here till Enoch was ready. He liked a captive audience.

Enoch started pacing, and North wondered how long the cult leader had spent watching TED Talks. Too long. 'The harvest of the seabirds is a tradition out on these islands dating back hundreds of years. I'd hazard men and women have hunted seabirds and their eggs for as long as they have fished and hunted animals on this island. We're honoured to have been part of that tradition. We eat together, we break bread together because we are family. Humanity has lived this long in this world partly by sharing food with one another. We in the Narrow Yett are no different.'

There was a cheer from the assembled crowds as Enoch looked out over the heads of the crowd. Enoch smiled – the kind of greedy, wolfish smile that made North shudder. He blinked and as he did the smile was gone and the mask back in place.

'And we will do the same in the next world.'

There was another cheer.

Behind him, North could hear a ruckus coming from the side room. Rowan was having problems. He caught Mia's soft voice at its most persuasive. If anyone could help Rowan get his sister, it was Mia. North wanted both kids and Mia out of this hall more than he had wanted anything for a long time.

'I am delighted to tell you that the time to move on is nearly upon us.' Enoch pointed upwards and around, and over the ceiling and every wall in the hall, a digital counter appeared. It showed 10:00 and a loud ticking noise started up as it began its countdown. Oohs and aahs from around the hall, the faces of his listeners lit not with a spotlight but with fervour and devotion. 'We who are the Narrow Yett are about to be transported by our Creators to a new simulation.' Enoch sounded thrilled at the prospect.

The crowd had begun to sway, their arms wrapped around each other. Across from North, a woman bowed her head and began to heave with ecstatic sobs.

'We who are the Narrow Yett will pass through to a better world. It will be our new reality. Our brother Jon Gaffney went before us and is waiting for us. Impatiently.' The digital clock went from the white plastered wall behind him. Enoch turned his back on his audience and spoke to the wall as if he could see Gaffney in it. Suddenly, green lasers lit up the plaster in the shape of an enormous gate, the serpent around it rearing up, and Enoch's audience gasped. 'Wait just a moment, Brother Jon.' He turned back to the crowd and raised his voice. 'Jonny was a beloved member of our community. He went into the World to tell humanity to pay heed to us – to warn them to be ready for the yett to open. And like that other faithful John. Like John the Baptist, he was martyred for his pains.'

Martyred? Gaffney had killed himself.

Eight minutes and thirty seconds left.

North started moving again, trying to get closer. He didn't have the time to find any stocks of cyanide and destroy it. There was time only to take Enoch down. He'd break his neck. Hold him tight against his own body with his left arm and bring round Enoch's head with his right. Bring it round too far and too quickly. Snap. He'd live with the consequences. Live with the fact this crowd would tear him limb from limb, but at least Enoch Fraser would be dead. He'd trust that Mia could figure a way out of this mess for herself, Rowan and the baby girl.

It was true to say Enoch loved the sound of his own voice and he was easing into a rhythm, feeding off the energy in the room. North had to hope that would give him enough time. He was helped by the crowd, which seemed to have lost all inhibitions. Moaning, some dropped to their knees with their hands raised into the air. Others reached out their arms to the gate, their palms upwards, eyes beginning to roll in their sockets as they sent up a low tuneless singing. And North realized they were not standing willy-nilly in the auditorium but were in fact snaked around the room in a queue, each of them with a small rucksack of belongings on their backs. It had to be why they were resisting him as he tried to push through. They thought there was some kind of order to whatever this was.

07:30.

'But I have even more shocking news, my children. The powers that be are sending soldiers to this island.' Gasps and mutterings. 'They intend to lock each and every one of us up. They will take our children from us. They will burn us from here and leave us bleeding in the streets. Above all things, they want to stop us passing through the yett. Because when this simulation quits, according to their programming – nothing is to be left of it. They want our annihilation.' Enoch was building. Passion entering his voice. 'But we who are the Narrow Yett know better. Through our transformation, when we leave this world, we will save humanity. Because when we pass through the yett, we will eventually create other simulations with other conscious creatures in them. And those conscious creatures will create their own simulations. Until the end of time. Our descendants will be the ones to create the very simulation in which we stand today. Together. Today, the laws of physics in this world are merely the boundaries on this simulation imposed by our Creators. We are our own Creators. And today, we rewrite our code, and in rewriting that code we empower ourselves to leave this dying simulation and pass through to a better one.'

North knew Enoch's conviction and the power of his delivery was more persuasive than his logic, which left him at least baffled.

04:00.

He noticed Comfort standing just behind Enoch on the stage. Her gaze slid over the crowd before coming back to him, and she shot him a look of vicious hatred. Turning her head, she whispered some aside to a heavy with ginger sideburns and his no-neck friend. They both started moving towards him. Ginger had a gun – a P226, North was willing to bet. Despite Drummond's best endeavours to keep his men away from the members of the Narrow Yett, a couple of his thugs had converted. That would be down to Comfort, he knew. She must have figured they'd come in useful.

He caught a glimpse of something being passed among those at the front of the crowd. Bread and red wine. He swore. 'No!' he called out, but no one was listening to him.

Enoch was in full voice. 'We must hurry through when the yett opens.'

The crowd chanted it back to him, swaying and glass-eyed. 'Open the Yett.'

03:00.

The two heavies Comfort had spoken to were suddenly on him, a searing pain as No Neck wrenched his arm up behind his back. Ginger pressed the gun into his ribcage. 'One more step towards Enoch and it's over for you, North,' he said, his breath stinking of rancid meat. North stopped moving.

'We have walked the paths of righteousness.' Enoch's voice was euphoric.

'Open the Yett!'

'We will rewrite our code and pass through the gate.'

'Open the Yett!' The crowd's chanting was growing in volume and passion. The atmosphere in the hall was getting to fever pitch.

Enoch spotted him. His finger pointed at North. 'Rewrite your code. Leave your pain in this world and join us in the infinite, Brother. Pass through the Narrow Yett with us to the Beyond.'

The crowd chanted, 'Open the Yett! Open the Yett!'

02:00.

'You're insane!' North raised his own voice and as he did, he brought his arm close to his body. No Neck behind him tipped forward as North pivoted his body towards him, using the heel of his hand to smash his nose, then punched him full force in his solar plexus. The man dropped to his knees. These two were Drummond's men. Not trained – just brutes who enjoyed violence. A trained soldier would have shot him. But a brute took an extra second to realize what was going on. And a second was all North needed to knock the gun up into the air, seize hold of Ginger's wrist, and use his boot to take out the guy's knee. Ginger shrieked like a banshee.

01:30.

North picked up the Sig Sauer he'd dropped. He raised the gun, but Enoch had come down from the stage to walk among the crowd. North shifted position. There was no clean shot. He had to have a clean shot. He raised his arm, firing into the air – one, two, three times. But there was no panic. No one dropped to the ground in abject fear. The shots didn't register. Bullets held no meaning for Enoch's followers in this world. He had to get closer to Enoch. He moved and as he did so he reached out and knocked wine and bread out of the hands of those around him. 'Wake the fuck up. He wants to kill you.'

Thirty seconds.

He kept moving, pushing, shoving.

Twenty seconds.

God, he hoped he was wrong. Prayed he was wrong.

Ten seconds.

Perhaps this was just a harmless ritual and the cyanide really had been meant for the rats?

The digital clock flipped to 00:00.

From the floor, the cult leader raised his arm and made a gesture to someone off stage, and the lasers flickered and shifted and the yett appeared to swing open. A massive cheer went up, wails and moans of delight. Almost as one they raised the wine to their lips, brought the bread up and started chewing.

'No!' North's shout went unheeded.

These people were too far gone. Too crazed with devotion, in denial about Enoch's intentions. Too mortgaged to the idea of salvation. It was too late.

There was a piercing scream as the first woman toppled to

the floor. Then another and another. Scream after scream now amid the chanting.

North glimpsed Enoch's face spasm into what he could only describe as visceral pleasure as his followers chewed and swallowed and drank down their cyanide-laced communions.

Some of the Narrow Yett at least appeared to have woken up to the danger they were in. They dropped the communion and started making for the door, only to clutch at their throats, fall to their knees and start writhing on the floor in agony. Others who attempted escape, driven away from the exit by the guardians with their bats and knives.

North had seen death before in war and in peacetime. He'd killed out of duty, in defence of himself and of others, and had killed in a spirit of revenge. He'd killed when his blood ran hot and when his temper ran cold. But this was something else again. A dark despair settled over him at the waste and stupidity and evil in the room. Enoch? He had to kill Enoch. But Enoch had gone.

Now all he could do was save the people who mattered to him. Mia and Rowan should be here by now. Where were they? What was happening in that side room? North broke through the crowd, smacking the beakers of wine and the bread from hands where he still could, stepping over writhing bodies where it was too late. He paused as he saw Ginger, his knee at a strange angle now, clutching his belly, bloody froth on his lips and his eyes rolling back in his skull. North bent down. 'You should have let me kill the bastard, pal.'

In the dark corridor, he could hear screaming and crying from the room where the children were kept. The door was locked. He lifted his foot and kicked the lock out, the wood splintering with the impact. Mia was on top of Rowan's mother. Rowan had his sister in his arms, the other children gathered round him, crying and wailing, the older ones holding babies. On the side were trays of juice and cookies, bottles of milk.

North picked Mia up from the woman and dumped her on her feet, then picked up the woman herself. Lisette spat in his face and attempted to claw at his eyes. He pushed her bodily to the far end of the room, opening up a closet and shoving her inside, before breaking up a chair and sliding a chair leg through the handles.

'I swear to God, that insane bint Lisette was about to poison the kids,' Mia said, gesturing first to Rowan's mother and then to an older straggle-haired woman already slumped and dead on the floor next to her. 'This one said the Narrow Yett was closing and she didn't want to miss her chance.'

North nodded. 'That's exactly what she was trying to do.'

North glanced over at the sobbing kids. Some wore small rucksacks. Others clutched hessian bags and teddy bears. One pale-faced girl in plaits hugged a rag doll to her chest. Their parents had packed for them, he thought. Their favourite toys to carry into the next world. They'd kissed them goodbye and let them go, trusting in the logic of a madman.

'Take the kids out the back door and get them out of here. There's help coming but I don't know how long it'll be. Get them to the tech hub. Barricade yourself into an office that doesn't have a window. Shoot anyone who tries to stop you getting out of here and shoot anyone who tries to get in to the kids. Can you do that?' He handed her the gun.

'Shoot someone to save the kids? Try stopping me,' she said, taking it. 'How do I use it?' Her eyes were huge. He forgot the impact a gun could have on someone who'd never seen one.

'Point, squeeze the trigger, and fire. Repeatedly. Do not hesitate.'

He shoved Mia's phone at Rowan. 'Redial the last number I called. My friend Fang will be on the other end. Tell her what's happening here and where you have the kids. Okay? I don't want them caught in any crossfire when help comes.'

'If' help comes, he thought.

Rowan stared at the cupboard where his mother was now imprisoned. Her shouts were muffled but the noise of her banging shook the wood, though it was built too solidly for it to come apart. North gripped Rowan's shoulder, feeling the fragile collarbone under his hand. The kid felt as if he was made of air and bones. And now he took the time to look closer, he could see a slap mark against Rowan's cheek where his mother must have hit him. 'It's going to be all right, kid.'

'She was going to kill Tara.' He held his baby sister tighter to him and she buried her face in his chest. Her arms were around his neck as if she was never letting go.

North nodded. 'But you stopped her.'

Rowan's face set. He turned his gaze on Mia. If he'd liked her before, he loved her now. She'd saved a room full of kids, including his sister. North got that, because he felt the exact same way.

North started moving towards the door.

'Where are you going?' Mia asked, a note of panic in her voice. 'Stay with us. We can wait this out.' And he remembered Fang a lifetime ago trying to persuade him to stay. He hadn't stayed then and he couldn't stay now. He had something he had to do.

'I need to find Enoch,' he said. 'Then I need to shove him through an effing wall.'

54

London

On a leather Chesterfield in a corner of the student union of King's College, Fang was bent over her laptop. She didn't think North would do as she'd advised. But at least he knew about the cyanide now and so did Hone. Would the navy have arrived yet? The army? Would the Americans from the spaceport? She was across Hone's computer monitoring events. But like she said, she could multitask. Frankly, she had no choice.

Her minutes of freedom were slipping from her grasp and she still wasn't done scratching the Benson Corporation itch in her brain. Fang firmly believed in getting her revenge in first. If it wasn't for Benson Corp, she wouldn't have Roxie on her case. And she wouldn't just have had to confess to Hone that she'd committed treason.

The one-eyed man had ordered security guards to throw her out of the building. It had all been very public. Fangfang Yu no longer worked for MI5.

She was still processing that conversation with Hone.

She was also moving every fifteen minutes to avoid Roxie. Not until this was done and North was safe was she letting the CIA agent catch up with her. Her body wanted to shake and tremble at the thought of Roxie. But she couldn't let it. She closed her eyes, opened them, moved the tip of her plait into her mouth and focused.

It was clear that the Benson Corporation had scant regard for laws or conventions when it came to weapons development for use in space. And that was fine with the US government. And indeed the British government. Fang took a second to wonder how many more companies like Benson there were out there. And if they were prepared to break laws and conventions when it came to the satellites, what else were they doing?

The memory stick, which Fang had given to Roxie, had been used in a computer located in the D'Urberville Technologies' London headquarters, which itself was routed into the wider Benson Corp system. Roxie had run a copy off herself before she passed it over to Mouse – thanks to which, Fang now had reserved passage through the CIA network, but she was saving that till later. For now, her attention was focused on Benson.

Whatever was going on was big. Laurie Sampson had been killed over it, and Lucy Sampson had only just escaped the same fate. The chances were the guy at the top of the company knew all about it then. Especially when you took into account the fact Mouse had served under him.

She filtered everything older than two years from Teddy's email system.

There was very little from Mouse. He and Teddy either had a different form of written communication or they handled everything on the phone.

She set up searches for anything that mentioned the Sampsons, Gaffney or Murdo. She hadn't been looking for Kincaid, but as she scrolled through Teddy's contacts, the name caught her eye. Kincaid, the guy who tried to kill North.

It was brief and to the point, Teddy asking Kincaid (whose first name had been Brad, she discovered) to liaise with Dr Libby Thornton and give her whatever she needed.

Dr Libby Thornton lived in Dallas, Texas. Benson Corp had a laboratory complex in Dallas, and Dr Thornton worked there.

Fang did a deep dive on the scientist. She was a molecular biologist with an impressive academic pedigree from Johns Hopkins. The Americans had lured her in, pretending to be someone they weren't. Well, she could do the same. Fang pulled up the members of the armed services committee and the names of their aides, sifted them for their politics. She set up an untraceable phone link and pulled up a voice synthesizer, thirty-something, male and American.

'This is Sam Whales, calling on behalf of Senator Rosie Johnson. Can you put me through to Dr Libby Thornton, please? Thank you so much.'

'Sam' bounced around Benson for a couple of minutes till a phone started to ring. A tired voice answered, 'Hello?'

'Dr Thornton. Hey, how are you? Senator Rosie Johnson of the US Senate Committee on Armed Services asked me to call you. Do you have a couple of minutes?'

The tired voice got peppier. 'Sure. Sam? Right?'

'Absolutely. Dr Thornton. Senator Johnson, as you undoubtedly know—' Implication: if she didn't, she should have. '—has a particular interest in your subject area. She's minded to bring you on in an advisory position. Dr Thornton, how would you feel about serving your country as an adviser to the defence committee?'

Social engineering was an art form, Fang always thought. A cocktail of tiny nuggets of fact blown up to appear like the whole truth, subtle intimidation, charm and flattery. Mercy for the person on the other end of the line had to be less than nil. Especially when they were in league with bad players.

Dr Thornton was wary. Fang could hear it in her voice. She probably worked in conditions of absolute secrecy, which meant, Fang knew, she didn't get to talk about her work in the way any expert desperately wanted to – at length and in great detail.

'Teddy Benson personally put forward your name to the senator, in fact. He speaks very highly of you. The senator and his mother, Mrs Benson, go back a long way.'

More than four thousand miles away, Fang could hear the stress ease from Dr Thornton. 'I see. Well, in that case, I'd be honoured.'

'To save us some time this end, Dr Thornton, I wonder if you could just run over the latest thinking in your field,' Fang said. 'And then the senator is particularly interested in what's going on in Scotland – "Murdo", is it?'

Dr Thornton started to talk, and Fang checked that the recording function on the audio file was working. Hone had fired her – everyone had heard him. All bets were off.

55

Murdo

In the refectory, the dead and dying were heaped on top of
each other. Their faces blue and contorted with agony. There
was still screaming and weeping, but it was quieter than it had
been. By the locked main door, bodies were piled up against
the wood. North unbolted it and pushed it open. Fresh air
swept through the hall. He wanted out of this hellish place,
but he still had to check in case he had miscalled it and Enoch's
corpse was in here.

A handful of Enoch's most devoted followers were patrolling
the hall. But they weren't providing ease or succour to their
friends – they were instead drawing a swift line across their
throats with kitchen knives. North thought about Gaffney.
These were his people.

North recognized Comfort's blue hair as she nursed Joy's
head in her lap. Her hands and the front of her jumper were
crimson.

She looked up as he came close. 'She didn't drink enough
of the wine.' Her tone was matter of fact. Any feelings of
resentment or dislike apparently forgotten in her focus on Joy.
'Such a scaredy-cat. I told her to drink it all down in one go.'

In her lap, Joy's eyes in the chalk-white and green-rose-
tattooed face appeared to plead with North. Terror. Pain.

Despair. Joy didn't want to die, and she no longer believed she was heading for any place but the grave. 'Comfort.' Joy's voice sounded like gravel. Her fist took hold of Comfort's jumper to pull her closer. 'Don't drink it. Promise me.'

Comfort lowered her face to Joy's and kissed her on the lips. She ran her pink tongue around her own lips as if to taste the poison there. 'See you on the other side, Beloved,' she said, and drew the knife across Joy's throat and then her own. An arc of crimson blood flew from her neck, and North felt it wet on his trousers.

'No!' He shoved her aside, reaching for Joy, pressing her bandaged hand against her throat with his own, and he could feel the pulse and push of her life force against it, but it was too little. He held Joy in his arms, a honeycomb of blood bubbles rising and popping at the corner of her mouth. A choking rattle in her cut-about throat. He felt the effort of her final breath shiver through her. Then her head dropped back in his arms and he realized she was gone. Green roses on a chalk-pale face, fresh red blood between them. Nothing to be done.

His hands bunched into fists.

Nothing except find Enoch and squeeze the breath from his body.

As he stood, he heard a choking noise from Comfort, stretched out face down on the floor, her fingers scrabbling to touch those of Joy. But she was too far away. He stepped over her.

The stage was empty now, other than half a dozen people who had flung themselves against the wall with its flickering green gate. Bloody handprints pressed against the plasterwork, reminding him of Gaffney's crude artwork in the cottage.

North did not believe that Enoch was fatally deluded. If that's all this was, he'd be dead along with his victims. Enoch was a psychopath and a serial killer. He knew he wasn't rewriting any code. Knew there was no gate to any other existence. This was the only reality there was and he'd chosen to stay living in it. There was no reason to preside over the deaths of his followers other than the gratification it gave him. He'd played a long con and when it fell apart, as he always knew it would, he'd turned to murder.

But this was an island and where would he go? Drummond wasn't waiting for his former friend. This whole venture had been about money for Drummond. Not power or religion. The Narrow Yett operation was a convenience for Drummond. A money-making machine and one moreover that allowed him to launder his own drug and crime monies. It had been useful as long as he could exercise influence over his old friend. When that went, the venture became more trouble than it was worth. If the police and taxman were going to start investigating the Narrow Yett, Drummond wanted out. Did he know what his old friend was planning? North thought he had known, which explained the haste with which he'd attempted to leave the island. He wanted nothing to do with Enoch's planned murder spree. He might even have thought he and his men would be at risk. Drummond had cleared out and left three hundred men, women and children behind to die. Including his baby daughter.

North wondered how much money he had washed through the Narrow Yett. A million? Two? Ten? Was that price enough for three hundred lives?

But he could track down Drummond later. Whether that was on the open sea or a tacky villa on the Costa del Sol. His target now was Enoch – the monster who had just persuaded hundreds of people to kill themselves. Who'd tried to kill two dozen children, including babes in arms.

There was a roaring above and the glass in the huge picture window shook in its frame. A helicopter hove into view. It turned and North glimpsed the shocked faces of the Americans from the bar staring through the glass and into the carnage.

The copter touched down on the grasslands above the settlement, then another. The American contractors spilled from the belly of the bird, fanning wide and all of them armed. Behind him, North heard the roar of SUVs approaching the settlement from over the hill. Hone must have decided to call in the Americans from the spaceport. It was a smart move.

A tall, thin man with bulging cheekbones approached. 'My name's Brett Munson. You can call me Mouse. Take this.' He handed North an earpiece. 'And catch me up.'

His last encounter with the Americans had not ended well. Kincaid had tried to kill him and ended up dead for his trouble. Mouse seemed to read the doubt in him.

'We're good, North. Your boss asked our government for help. They asked us to turn out. I'm required to provide all reasonable assistance. Your own people will be here soon.'

North's eye snagged on a small camera in Mouse's vest. He was broadcasting back to North didn't know where, but he had to presume it was a live feed. There were some advantages then in working for a satellite company.

Benson wouldn't be his first choice for the cavalry. But these were extreme circumstances. He nodded. 'The adults are dead. Mass suicide. Their leader is a guy called Enoch. And he's not among the dead. I need your guys to check for survivors. See if there's anything to be done for them. It's possible there were hold-outs who are still in the cottages. But the priority is that place.' He pointed and Mouse turned and whistled, sweeping an arm towards the refectory. As one, his crew started moving, medics among them. North hoped they knew what to do in the event of cyanide poisoning. 'We have at least two dozen children safe in the tech hub. They'll need bringing out and taking off the island as soon as possible.'

Mouse put a long skinny finger to his own earpiece, as if being questioned down the line. 'And a guy called Drummond? Apparently, your boss wants to know where he is.' Mouse's voice was loud over the still-moving rotor blades. 'Is he among the dead?' So Hone was being patched in live to this, North realized.

He shook his head. 'Drummond and his men took off by boat. I don't think he wanted any part of what was going on here.' So much so, he was willing to let his baby daughter and the woman who passed for his wife die with everyone else.

Frowning, Mouse turned to gaze out at the sea. 'It's rough out there. High winds. They may not make it.' Which was good news. As far as North was concerned, Drummond at the bottom of the ocean was ideal. Mouse, however, had different ideas. 'We'll recce this, and send someone out to pick them up.'

'I just want Enoch,' North said – the thought of Comfort's rank stupidity, her malevolence, which lacked all conscience, and most of all Joy's death in his arms still fresh in his mind. The hot, wet crimson blood soaking them. The terror in her eyes. That final wracking breath. 'It's an island. There are only so many places to hide. Can you spare two men to help search?'

Mouse opened his mouth to reply, but as he did a shot rang out.

Smoke was rising from the tech hub. North started running.

56

There were at least two dozen children in there, as well as Mia. And he wasn't losing Mia, not again. This time he would be there for her, and nothing and no one was getting in the way of that. Certainly not a murderous megalomaniac like Enoch.

By the time he made it to the hub, smoke was already curling out from under the ancient oak door and from under the eaves of the tiled roof. As he pushed open the church door, he heard shouts from behind him. He didn't know whether it was Mouse telling him to wait, or the contractors alerting each other to the fire. But they had enough on with the dead and dying and he couldn't wait – this was his fault and he was the one who had to sort it. His heart thundered in his chest – was this what panic felt like? He'd told Mia to lock herself and the kids into the tech hub because he thought they'd be safe there. He'd second-guessed Enoch and second-guessed him wrong, presuming he would hole up in the manse or even head for the quayside in a bid to escape the island. But either of those options would mean Enoch would have to stop talking, and above all things, he thirsted for an audience.

He made out a tinny voice a million miles away.

Hone. He dug his own earpiece out of his pocket and screwed it into his ear, as he crossed the stone flags at a run heading for the glass door into the main atrium, behind which he could make out billowing smoke. He seized the handle and pulled but it didn't move. North cursed. There was a keypad by the door, blinking red. It was locked.

'Enoch has called into a radio station,' Hone said. 'We're trying to persuade them to pull the call, but it's going to take a minute.'

North looked around. They'd kept the space in the church as a rec room with old leather sofas and coffee tables made of crates. It was sparse but that was okay, he didn't need soft furnishings right now. He needed something heavy.

Hone was obviously listening to Enoch's broadcast while still on the line to him. 'Enoch's telling the world...' – pause, while he heard Enoch out – '... that British and American troops have slaughtered his followers...' – pause – '... and that they've set fire to the building he's standing in, which is...' – pause – '... full of innocent children. He said he's trying to save them.'

At the mention of children, North heaved a Chesterfield out from its space, turning it lengthways. Thank God it was on casters. The kids were beyond that door, locked in a burning building with a mass murderer. 'Where's Fang?' he said.

Hone took a beat. 'Busy.'

'Have you heard from Mia?'

'No. Focus, North. Enoch is saying he wants humanity to know the truth about reality. That this isn't it. That everything around us is a fake. That governments don't want us to know the truth.' Truth? Enoch didn't know the meaning of the word. North put his hands to the sofa, squared his shoulders, and started to run. 'The presenter knows something's going on. He won't shut it down.' Hone in his ear, where Fang should be. Busy? What did 'busy' mean?

As the end of the sofa struck the glass, the door shattered into a thousand pieces, the shards bouncing and smashing into a thousand more against the stone flags. Smoke billowed out from the open doorway. 'That bastard,' North said between coughs as he hurtled over the sofa, 'just presided over a massacre here. He's done enough damage.'

North could see the flames from where he stood.

'Where is he, Hone? Any clues in what he's saying?' He stripped off his hoodie and bundled it into a mask in front of his face, but he started coughing anyway as the hot smoke forced its way down his throat and into his lungs.

Hone replied but he couldn't make out the words. He'd find him.

'Mia!' He lifted the cloth from his mouth and attempted to call out, but the ever thickening smoke seemed to swallow the word. Somewhere to the left of him, in what he remembered was an edit suite, a computer exploded in the heat, and then another to the right, and another further down the corridor. He attempted to get his bearings. He was on the ground floor, standing in a corridor with edit suites and a small studio branching off from the main corridor. A corridor that led into the main atrium.

He'd told Mia to head for an office that had no windows. The offices were all upstairs on the first floor. The natural choice would be the office at the furthest end of the corridor. He'd been relying on the fact he could get to her via a long hallway to the right and then a staircase, but even as he took his first step along it, there was a roar as the ceiling came down belching black smoke and flames, blocking the way through. That meant he'd have to go through the main hub and he was willing to bet that was the seat of the fire.

There was another gunshot. Was it to let him know she was in trouble, or because Enoch was attempting to breach the door? Surely Mia would have moved the desk and filing cabinets up against the door? But if she could smell smoke, she might be planning to make a run for it with the children. And in that case the shots were to lead him to her. The problem was, North wasn't the only one hearing them.

He called out her name again, his voice cracking. He wanted to think that was down to the smoke. But he knew it was from fear – not for himself, but for Mia. There was no reply.

'... reality...' – North caught – '... simulation.' At first, he thought Enoch was in front of him. But then he realized Hone had patched him into the radio broadcast.

His face buried in the crook of his arm, he staggered onwards, Enoch's voice fighting against the crackle and hiss of the fire. 'The Narrow Yett has opened. Be ready to walk through...' As he pushed open the door, for a brief second he thought he saw Enoch's face on the enormous screen as if he had willed himself there. But then the screen went black as flames burned through the wires. And the radio broadcast ceased. Hone must have forced them to take Enoch off air before he had the chance to call on listeners to commit suicide and murder.

'Mia!' North shouted as he went. 'Mia!' He knew where she was but he wanted her to know he was here.

A computer exploded in the heat and then another and another. The enormous screen burning through from the inside out, loading the air with noxious gases. The heat building second by second. The tech hub was turning into an oven.

'You ruined everything.' Enoch's voice sounded so close, he could have been standing in front of him. It took him a moment to readjust – it wasn't the radio broadcast any more. It was a flesh-and-blood man. North reached for Enoch with outstretched arms but met only thin air. He cursed. He should have demanded a gun from Mouse instead of running straight into the burning building.

'You don't understand, North.' From the right.

'They had to die.' From the left.

'They all had to die in order for them to live again.'

Disgust and rage flared in North as he turned this way and that in an attempt to figure out where Enoch was. The guy was older, less substantial. He should be able to overpower him within seconds – if he could get his hands on him. He tamped down his frustration. It wouldn't help him. This was Enoch's home ground. The man might look mild and harmless, but he'd just proved himself capable of mass murder. He was a cold and vicious predator.

'You should have put aside your pain and started over with the others.' That rich, mellow reasonable voice, taunting him.

North made a grab, but Enoch was too fast for him.

'But of course you were never capable of that, because you're a virus in the system!' North felt the air next to him shift and caught a glimpse of a shining blade. Then Enoch was in front of him, six feet away, an enormous axe in his hand and his eyes insane with fury.

'You got what you wanted, Enoch.' North considered his options. Overpowering a man with an axe was a different proposition to hand-to-hand combat. The axe was sharp, it was dangerous, and it gave Enoch a reach he didn't otherwise have. 'They're all dead, aren't they. Your followers. Job done.'

Smoke billowed between them and when it cleared Enoch and his axe were gone. 'So why don't you let me help you through that "yett" you're so fond of.' North listened hard, rotating, tiny step by tiny step, the hairs rising on his neck. Where was he? 'They must be waiting for you in the next simulation. Right?' A breath of air. He shifted his weight and flung himself against where he figured Enoch was, but still there was nothing. A peal of hysterical laughter came out of the smoke ahead of him and then what felt like an office chair slammed into him and drove him backwards so fast his feet went from under him; his head caught the corner of a desk and then slammed into the floor. He had a second to think about the bullet in his head, willing it to stay where it was, before rolling to one side just as the blade of the axe bit into the concrete floor next to him. Enoch screamed in frustration and North scrambled to his feet, coughing. The smoke was in his lungs now. He could feel it spiralling and writhing in his chest. It didn't seem to bother Enoch. 'You told Gaffney to make a sacrifice on Tower Bridge, didn't you?' Enoch loved an audience. If he could get him talking, perhaps he could distract him long enough to bring him down.

Enoch cackled. In the shadows and the smoke again. Like some devil out of hell, North had time to think. Not a man at all. 'I had to get their attention. I needed to let the world know that the Narrow Yett was about to open. Pain and suffering mean nothing. Because this isn't reality. This is an illusion of reality. Once you accept that truth, anything is possible.' The voice moved as Enoch continued to speak to him, circling him in the darkness and the smoke. 'Those dead people you tried so hard to keep alive, they're only so much coding in this world. But in the next they live again.'

'Then why aren't you with them? Why stay?'

'My mission here isn't done yet,' Enoch said in a snarl.

'Drummond thinks it's done. Did he have enough of this madness?' North was looking for an opening.

'I never wanted Drummond here. He gave me no choice. He said I owed him.'

'Owed him?'

'For taking care of Maria.'

Maria? His dead wife. The wife who had fallen and died skiing off-piste.

'Drummond murdered your wife for you?'

'She had so much money and she was such a dull woman. So very ordinary.' Ten years ago, he had conspired to murder one woman. Today, he was responsible for the agonizing deaths of three hundred. Enoch had come a long way.

The air moved, and North realized Enoch was behind him. Turning, he rushed him, his arms wide, but instead of catching hold of Enoch, he tumbled over a desk, toppling head first on to the floor. Pain and stars. Enoch tipped the desk and it hit his chest and temple with a glancing blow as it continued its somersault across his recumbent body. North closed his eyes and readied himself for the bite of the blade.

Another shot rang out.

Mia.

He took a breath of acrid air and opened them again.

Flames suddenly lit up the smoke, which was growing more noxious by the second. Enoch was gone and with him the axe.

Head ringing, chest singing with pain, North staggered to his feet. Which way?

Then the children's screaming started. Thud – scream. Thud – scream. Enoch, who knew the building backwards, had found where the children were hiding. Struggling to remain conscious, sound coming and going, North started running, bouncing from wall to wall as he fought to keep upright. He needed to make it to the staircase.

'Suffer little children to come unto me.' Enoch's face was pressed against the closed door as North took the stairs three at a time. Even from where he stood, North could see the wood was splintered around the lock. It would take one blow more and the lock would give. The work of a second. He willed his limbs to move faster. He would be too late to stop Enoch swinging the axe. He would kill Mia with the first blow. She would be standing in front of the children, using her own body as a shield, he knew that. With Mia dead, how many children could he kill between now and the moment North took him down?

The final few steps and North opened his mouth to yell. To distract Enoch for a second. For two. Mia was about to die. But there'd be one less swing of the axe. One less child hurt. Willing himself to move faster.

And as the lock failed and Enoch took a step forward, there was the sound of a gun firing. Another shot. And another. Enoch staggered back one step, two. And as he did, Mia emerged from the room, face grim, arms outstretched, still firing the P226. And she was still firing it as Enoch tipped over the metal banister. Bang. Still firing it as his arms and his legs rose together. Bang. As his wide staring eyes grew dull. Bang. And he dropped into the inferno below.

57

London

The Americans offered a live feed and MI5 took it. Hone sat in his office and watched the tech hub burn. The contractors were doing their best to calm down the flames, but it was an impossible fight and one they were losing.

In a coffee shop a mile away, Fang watched the video from Murdo even as she sent the audio recording of molecular biologist Dr Libby Thornton to Bald Paulie with Lestat's contact details. Paulie would know what to do. And Lestat would help him.

The pictures from Murdo were jerky. Handheld and not by a professional. Others from wearable cameras. A dozen adults had survived. They sat apart from each other. Survival blankets around them. Some of them rocking backwards and forwards. One of the cameras moved closer to offer what appeared to be a bottle of mineral water. All the survivors wore the long-distance stare of someone in deep shock. Half a dozen corpses were laid in a neat row on the ground. They were the ones who had staggered or crawled outside to die and had been cleared to allow the Americans access to the building. A total of seventeen members of the Narrow Yett had been triaged in situ – none had survived. Mouse had ordered three separate counts of the bodies – 288 deaths. The rest left in the refectory where they'd died. British forces were understood to be disembarking at the quayside.

Initial reports from the US distilled the scene into one word, and that word was 'Carnage'.

But Fang couldn't think about any of that.

Mouse reported the tech hub was burning.

There were children in the building that was ablaze, and Fang knew North. He wasn't coming out without them.

In his office, Hone cupped his hands around the flame of a match and raised it to the cigarette between his lips. 'It's been too long,' he said. He spoke as if to an empty room. He knew then, she thought, that she was watching these pictures with him, that she could see him in a small box on her screen, that she could hear him.

Fang planted her glittering feet more solidly on the ground, sat forward in her seat, and pulled her leather jacket more firmly around her.

'And this is a lot of attention for one small island,' Hone said, as if reflecting on the weather. 'Your American friends aren't going to like it.' He looked directly down the camera lens.

He was right. And when news of the mass suicide broke, the world's media would want a piece of the action and anything and anyone on the island would come in for the most intense scrutiny. Whoever was left at the spaceport must already be packing their bags, she thought. After this, all dubious activities involving militarized satellites would cease – at least on Murdo.

The spaceport would go back to being a genuine commercial hub for communications satellites or even a simple rocket spotting station. The Americans might give up their lease rather than be even tangentially associated with a mass suicide.

And even as she thought it, through the heat haze on screen she thought she glimpsed three US helicopters fly overhead and out to sea.

Right this second, she didn't care. Not about any of it. She cared about one thing.

Don't let him die. Her fingers hurt, she had them clenched so hard.

Hone had smoked his cigarette to the filter tip. He ground it into an ashtray on his desk and let go a breath and as he did so she thought he let go of North. He drew up his chair to the screen. 'I've things I need to do. And so do you.'

Life without North.

Such a moron-person, she thought. Her hands went to the top of her head and she squeezed her elbows together. Desolation. Her friend. Her brother in arms. This was always how it was going to end. Why hadn't he just stayed in hospital as Linklater told him to?

This whole mess was all Hone's fault.

She typed one word. 'Catina.'

Hone looked blank.

She kept typing. The text appearing and disappearing almost as quickly. 'Catina. The girl who was sex trafficked. She stays. She goes to college. Find her somewhere to live.'

Hone hesitated, then gave a cursory nod to the camera.

'Now, feck the feck off my system, Fang. I fired you, remember.'

58

Murdo

Hundreds of miles away there was the banging of bolts and then the sound of an oak door opening. Smoke billowed from the doorway as from the gates of hell, and a stream of kids ran out screaming, then Rowan emerged carrying Tara, then Mia, kids in either hand, and finally North, festooned with children. Two children on his back, a child in his right arm, two more standing on his shoes and clutching his thighs. He half carried and half dragged them out into the fresh air and away from the tech hub, flames surging from the roof of the building now as it shrieked and screamed in the infernal heat.

The children still clinging to him, North turned round. The building was swaying, as if in two minds as to whether it wanted to stay standing. With what sounded like a heavy sigh, it let go and crashed to the ground in an explosion of flames and steel, a huge cloud of smoke and dust rising into a cloud above it. Enoch was gone. Buried in the ruins of his own delusions. And good riddance.

The Americans came and removed the kids. The small faces stricken and terrified. North watched as Mouse shouted orders to keep the kids away from the bodies on the lawn. 'Get them triaged at the manse and given water,' he called out.

'Hey, North,' Mouse was calling to him now. 'I hear we owe you a lift.' Mouse pointed behind the settlement.

The Americans came over the horizon like so many Iron Men, some in jet suits, others on hoverboards. Want, he thought as the first pilot came down a few yards from him, settling on his feet, the powerful engines loud even over the noise of the sea and the fire still raging in what was left of the tech hub.

'Tell me you have one of those for me.' He gestured to the suit.

Drummond's Zodiacs hadn't got far. Despite the waves, they were doing their best to stay in some kind of formation. They started firing as soon as they glimpsed their pursuers. One of the Americans on a hoverboard was shot, the board spinning him high up into the air before it crashed back down into the sea. North swerved, firing his own gun and taking the shooter in the boat clean through the forehead. Drummond stood in the rock and sway of the boat gazing up at the flying men around him, his attention moving from one to the other, apparently searching for someone. Their eyes locked as Drummond found what he was looking for. He levelled his machine gun and started firing at North, his face a rictus of hatred.

The Americans had given him pointers about flying. 'Don't scratch your nose,' the big man from the bar had said, pointing to the fire emerging from the turbine mounts on his arms. But North was still no expert. His heart pumping, he dipped and swerved, dropping so low his heels grazed the surface of the water and sent spray up into the air, his own shoulder-mounted weapon spitting out bullets so fast the action reverberated through his body. Drummond was a malevolent force. Whatever other crimes the gangster was responsible for, he could have stopped the massacre on Murdo. Instead, he had cleared out. He was as guilty as Enoch for the deaths of nearly three hundred people.

With a heart-pounding rush, North swept back up into the air, banked, and readied himself – he didn't need the gun, he was his own weapon.

This was it.

He headed for Drummond, watching the Glaswegian gangster's face shift and alter as he approached at a rate of knots. He felt the shuddering impact as his heels went into the other man's chest, knocking him off balance. Felt the tangle of their bodies together in the rubberized boat. Drummond let go a massive punch that would have broken his jaw had a wave not hit the Zodiac at that precise moment. North swept up his arm with its turbine engine, meaning to knock Drummond clear out of the boat, but the momentum of it took him out too, the icy cold water taking his breath away as the pair of them crashed into the sea and plunged downwards.

North looked upwards. The bottom of the Zodiacs visible, the spit of bullets into the sea from the Americans as they took down Drummond's remaining men. He waited for the pull of a lifejacket to rip him back up to the surface. Then realized it wasn't coming. The big man from the bar had disabled it.

A moment of panic.

The Glaswegian gangster hadn't spent the last minutes carrying Enoch's victims from a blazing inferno. North was tired and it was beginning to show. But more than that, he was weighed down by the technology he had strapped to him. He struggled for the emergency twist release, but his fingers slipped from it as Drummond found him again in the water. The huge hands closed around North's throat and began to squeeze, the edges of North's world turning dark as he stared into Drummond's face, the narrow eyes crimson with broken capillaries and the desire to kill someone – particularly him. North slammed the engine strapped to his right arm into Drummond's hip.

It wasn't enough to make Drummond let go, but the gangster, too, needed air. He surged upwards, North keeping tight hold, and they broke free of the water together, gasping, retching and coughing up salt water. But North could feel the weight on his back beginning to take him down again. Finally, his hands found the twist release and he felt the suit detach from him.

'I'm going to kill you.' Drummond was on him again, shaking him bodily in the water as if he couldn't decide whether to drown him or strangle him. The urge for murder so intense, he appeared to decide on both as he started to squeeze while attempting to push him under. But if Drummond pushed him under, that would be it, North knew. It wasn't happening because it couldn't happen. Because if it did happen, he was dead. And Drummond knew it too.

His mouth clamped tight against the slap and swell of water, he reached up to Drummond's hands. The gangster gripped him more tightly as if this time he was never letting go. North's vision began to close in on itself. Fighting for breath, he took hold of Drummond's little fingers and bent them backwards as far as they could go. Even with the sound of the sea and the wind in his ears, he heard the 'snap snap' of it a second before Drummond's yelp. And as he did, he felt something enormous and hard against his back. It slid up and along him, the force of the displaced water sucking him down as it rose behind him.

Drummond realized the danger before North. But then Drummond could see the Royal Navy fleet submarine emerging from the waves. North could only feel it. Drummond let go, launched himself through the water, and attempted to swim away to safety with a ferociously powerful crawl. But the backwash was sucking him under. His mouth opened in a scream as he disappeared under the curved steel body of the sub. North bumped and rolled down to the fin, water sluicing its curved sides, grabbing and wrapping his arms around the smooth metal. If he let go, he was dead. And with a final push, the submarine made it to the surface in all its immense metallic glory.

North sucked down the fresh air, the water still pouring from the surface of the submarine. There was a shriek of steel against steel from somewhere above him and the sound of grinding spinning metal. Blinking, salt water in his eyes, North looked up. From the bridge, a familiar face stared down at him. Whiskers. A jaunty black cap. Mac spoke first.

'The captain figured I'd know where to find your pal with the blow-up boats. I said you should have come back with me, laddie.' He winked.

59

New York

In a boardroom that looked over the city skyline and a glorious New York morning, nineteen men and seven women sat around a highly polished table. Many of the Benson directors ranked on the Forbes Rich List, and not just Teddy – five of them had service records in various branches of the US armed services. It was an emergency meeting, but the Benson directors appeared relaxed. After all, this was not the first emergency meeting they had held. And it wouldn't be the last. Only one face was missing – Mrs Benson, whose attendance was via a video link.

Teddy had the chair. 'Ladies and gentlemen, we come to the vote. A decision was taken under Benson's 2019 emergency protocol 267. The Murdo spaceport and its solid laser weaponry initiative into low Earth orbit has been shut down.' He gave a dry cough and reached for the water glass. His eyes flickered to the video screen where his mother sat, her face stone-cold beautiful and entirely impassive. Mother didn't like failure. He tried not to think about the death of Dr Laurie Sampson, which he'd ordered at the very start of this. Tried not to remember the failed attempt against the life of the engineer's sister, Lucy Sampson. But what a horror show Murdo turned out to be. No one could have predicted a mass suicide. And the general public didn't know the half of it. 'All sensitive equipment is being boxed and shipped out as we speak. We'd like to offer formal thanks to Admiral William Costas of the US Navy for diverting the aircraft carrier USS *Powell* to the island to help in the shutdown and clean-up effort.' Teddy nodded to a man in a dark suit, who gave a small shrug as if to say it was nothing. 'Under our standing orders, we require a vote by the full board to endorse the decision. All those in favour of the decision taken under the emergency protocol 267 please raise your hands.'

The show of hands was unanimous, and Teddy glanced again at the screen image of Mrs Benson. He'd tried to persuade her to abandon the entire project. He'd never felt entirely comfortable with it. Ratcheting up weaponry in space was a dangerous game. But he'd failed to convince her. She was resolute on this, as in all things. Which was beginning to irritate.

'Excellent.' Mrs Benson leaned closer to the screen. Her make-up was flawless, although the screen magnified the powder caught in the crevasses and there was little to be done to stop the bleeding of her carmine lipstick into the wrinkles around her mouth. It crossed Teddy's mind that he'd noticed this morning, for the first time, that his mother was perhaps wearing a honey-blonde wig. It suited her. 'We take comfort in the fact we have three alternative locations available to us here in the US and five abroad.' Mrs Benson's arctic blue eyes flashed with satisfaction. 'I see no reason why our space programme has to be put back.'

Her listeners slapped the table with their palms in applause. 'The potential risks to the security of the programme are being addressed and we are full steam ahead.' Again, the applause.

★ ★ ★

And in a long, low building that went unmarked on any map of Murdo's spaceport, a knife-thin man whose shoe squeaked closed a manila folder. It was late and the thin man was tired. It had been a difficult day. A day full of corpses and anguish, which he knew would stay with him for a long time. Which may be why his grip slipped and the folder fell to the floor. He cursed as he gathered up the pieces of paper and the photographs. Allowing himself a second to gaze at the photo of a bespectacled man the photo identified as being Dr Laurie Sampson, the word 'Deceased' stamped across it; accounts of the visits and habits of renowned space anthropologist Dr Lucy Sampson, marked 'No Further Action'; and a file within a file of snarling teenager Fangfang Yu, a yellow Post-it note with the words 'Recruitment Target' fluttering under the desk. He made a face no one could see. He regarded his work as a vocation and he was good at it. Laurie Sampson had been textbook. Lucy had proved something of a professional embarrassment, but he hoped she was allowed to live out her days in peace. As for Fangfang Yu – he loved her, not least because she was his passport home. And once she was stateside, Mrs Benson wanted a close eye kept on Little Miss Yu and he'd been happy to volunteer. He had to hand it to the Bloody Widow, she had great instincts when it came to potential trouble.

He reported directly to her. Sure, he'd served under Teddy and he followed Teddy's orders and one day the guy would take over. But until he did, Mouse ran everything by the Old Lady. Mouse didn't envy Teddy if he ever decided he wanted to do something a different way to his mother. Because that was never going to happen. Teddy might well get to be President, but it would be Mother who was running the country. And he, Brett 'the Mouse' Munson, would be right next to her.

He flipped open another box that bulged with files. In it

there were photographs, biographies and the medical records of every member of the Narrow Yett. With special attention paid to those souls who had killed or mutilated themselves within the last year. The bodies of the three suicides had in fact already been dug up from the graveyard and their remains shipped out for cremation. Gaffney, he knew for a fact, had already been cremated and his ashes disposed of. Mouse paused as he regarded the photograph of Enoch. Enoch Fraser had always been a psychopathic lunatic. Benson Corporation could not be held responsible for such a freak of nature.

What was equally certain was that it had been a mistake for Teddy Benson to instruct his man Kincaid to cooperate with Dr Libby Thornton. But Dr Thornton's evidence had been compelling. She had modelled the aerosol neurotoxic bioweapon thoroughly. She had experimented on mice, rats, rabbits, beagles and then primates of all kinds. She was convinced her bioweapon rendered subjects relaxed and cooperative. That in some instances they appeared to show signs of complete contentment, approaching a state of bliss. It was potentially transformative, she argued. An occupying army could sweep in and take over a city whose citizens would welcome them with open arms. And Teddy Benson had bitten.

All they needed was willing subjects.

Or rather subjects.

Murdo seemed perfect.

Kincaid had links with Drummond, who ran the place. He'd pitched it during one of their poker games and Drummond had signed off on it. Told Kincaid to wait till he and his own men were off the island. Demanded £250,000, and the helicopters went over the settlement the next night. Thornton flew in for it all. They'd set up CCTV in key locations and at first it seemed to work on men and women the same way as it had on the animals – feelings of relaxation that in some cases blossomed into euphoria. People hugged and danced but it unwound fast as the Narrow Yett followers lost all inhibitions. Sex happened anywhere and everywhere. Fights too. They sang, they screamed, they walked in circles till they dropped and passed out cold where they fell. Benson's people went in with biohazard suits and respirators. Under Thornton's supervision, they took blood tests and ran condition checks on all, then ripped out the cameras. Most subjects woke up the next morning with a sore head and no memory of the night before. But a handful woke to paranoia and derangement.

Over the next month, there were three suicides and five mutilations. Drummond demanded more cash when he realized they were as a result of the experiment. More again, when Thornton wanted those who had been most affected regularly monitored. She blamed a combination of the personality types of the subjects – vulnerable, naive, prone to hysteria and group-think – and DNA. Anything but blame her science. Every month for a year, Drummond's men had made sure the five Narrow Yett faithful were drugged and escorted to the bio-lab for tests. She'd been beyond furious when she realized Gaffney had been allowed to leave the island, sending Benson's men out to find him. But they failed.

There was no sign of the bioweapon now, of course. Not a fibre or a hair belonging to the Narrow Yett cult members remained in the lab. Mouse switched off the lights as he himself carried out the last cardboard box. The laboratory would stand as an empty building within an empty compound until it fell into ruins, which was, he decided, no bad thing. Thornton was still tweaking the science. She declared herself confident.

60

London

She had to go home sometime. North was safe. Lucy was safe. Bald Paulie had what he needed.

It didn't take long to pack her bag. She kissed her dozing grandmother on the cheek then let herself out of the back door and into the yard. She'd known she would feel bad, but it was worse than she'd been expecting. As Roxie had promised, the car was waiting, and the agent stood leaning against it, a bored expression on her face.

'Ready for the Land of the Free?' Roxie said and smiled, the milk teeth making her face more sinister rather than less.

Fang would have been fine but for Plug.

'Where do you think you're going?' he said, his huge bulk looming out of the shadows, his hand on her shoulder.

Two men in dark suits got out of the 4x4. 'Fangfang Yu,' one of them said, the accent unmistakeably American. 'Please come with us.'

Out of the corner of her eye, Fang sensed Roxie slide a pair of sunglasses down from the top of her head. Her hand slid under her jacket and went to what Fang knew was a gun.

'It's all right, Plug.' Fang tried to keep her voice nonchalant. Plug blocked her path to the car.

'This must be the famous Roxie, I'm guessing. Aren't you the Big I Am, bullying little girls.'

'Stand down, Mr Donne.' Roxie's voice was chilly. 'This is not a fight you're going to win.'

Plug's scowl would have stopped a tank in its tracks. He kept his eyes fixed on Roxie. 'Go back in the house, Fang. You're not going anywhere with this woman.' Plug turned to face the two men. 'Oi, goon squad. You might want to feck the feck off with the wicked witch of the west here. Before someone gets hurt. And when I say someone, I do mean you.'

Fang tugged on his sleeve but he ignored her, even as the two men stepped towards him. 'Plug, I have to go with them.'

He still didn't look at her because that would have meant taking his eyes off the Americans, and she realized he had knuckledusters on the fists of both hands. 'You don't have to do anything, Fang. Call that one-eyed bastard. Get us some backup.'

Fang stamped her foot. 'There is no "us". That's over, Plug. You don't understand.'

Across the road, a curtain twitched. Someone was already calling the police. But that wouldn't help, because there was nothing to be done about it.

'She's right, Mr Donne,' Roxie said, opening the passenger door. 'There is no "you" now – it's us. We're her new family. Get in the car, Fang.'

Another SUV pulled up behind the first one and four more agents got out. One of them with a neck as wide as his shoulders. At an unseen signal all six rushed Plug, and Fang heard the crack of metal against jaw. They were taking no chances. They knew who Plug was and his history as a cage-fighter. Plug was getting some heavy hits in, but there were too many of them.

'Stop it,' she was screaming now. 'Don't hurt him.' At some unseen signal, the agents stood away and one of them tasered Plug. Plug still stood. Another tasered him. And another, and Plug fell to his knees, then keeled over on one side.

'Stop it,' Fang screamed. 'Stop it.'

Plug was convulsing. His eyes turned up in his skull. And still they kept up the tasering. Fang tried to go to him, but unseen hands held her back.

'The only way to stop it is to get in the car.' Roxie was next to her, her eyes fixed on Plug's convulsing form. She was smiling.

'Okay, just stop, please.'

Roxie raised a hand and the agents released their triggers.

Fang felt the pressure of Roxie's tiny hand in the small of her back, pushing her towards the car. As she glanced back, she saw the agents closing in on Plug's body. One leaned down, she guessed to free the barbs from her friend. And as he stood back, he kicked Plug in the spine. Hard. Dispassionate. As if it was the next thing on his To-Do list. She screamed. The other agents began to do the same, working their way up and down his body. She was still screaming as they dragged her into the car. 'He hasn't done anything. Leave him alone.' Then the door closed on her. She flung herself against the back window, her hands pressed against it, watching the sight of Plug getting battered by the CIA agents getting smaller and smaller as they drove her away.

61

Murdo

The fleet sub was out of Faslane on the Clyde and had been on exercise when the emergency call went out from MI5. The captain would have taken him back to the mainland but North preferred to get back to Murdo. A crew member lent him dry clothes. As he stepped back on to the quayside, it was a different scene again to the one he'd left. The island swarming with Scottish police.

Rowan refused to let go of his baby sister till Mia persuaded him to let the medic check him over. Only when she held out her arms for the little girl did Rowan submit to the examination. But even so, his gaze remained fixed on Tara as if she might disappear if he stopped looking at her. Mia held the little girl on one hip and her grip tightened as Rowan's mother was escorted past them and onto a waiting boat.

Lisette acknowledged neither of her children. Her eyes were glassy. North wondered what she had thought as she'd been led through the hall full of corpses of those who'd once been her friends. Whether she regretted the fact that she had missed her chance to die and make it through to another simulation. Or whether she realized she had only just escaped being played for a fool. She was lucky to be alive, he thought. And only thanks to Mia and Rowan were the children she had had in her care lucky to be here too. In a quiet voice, North told the officer in charge that the woman was not a victim. That she'd tried her level best to kill each and every one of the island's children.

They planned to take Mia and Rowan and the baby off on a separate helicopter to the other members of the cult. Those who had changed their minds at the last minute. As she stood in the lee of the cottage that had been her home for five months, Mia held the gun out to him.

'I think this is more your style,' she said.

'I don't know,' North said. 'You seem pretty at home with it.' He took the gun and tucked it into his waistband.

'I killed two men today, North.' Her face was serious. 'Kincaid and now Enoch.'

'They weren't nice people,' he said.

With a sad smile, she reached for his hand. She glanced over her shoulder to make sure Rowan and his sister were still together. 'Rowan's asked if he can come live with me. The baby too.' Mia's voice was strained.

'That makes sense. I wouldn't let his mother keep a cat. And anyway, I think she'll be spending the next few decades in prison.' He drank in the sight of Mia's face while he still could. The dark eyes. The beautiful mouth.

'I don't want to be a mother. I shouldn't be anyone's mother. I don't know how.' The dark eyes were fearful. The beautiful mouth turned down. They both knew what she meant. That her own mother had beaten her too many times to count. That she'd starved her and locked her in dark places. That there were scars not even time could heal.

'Rowan's no fool,' North said. When he knew what he

should say was that Mia was brave and capable and loving. That she was generous and strong. That he knew she would already fight tooth and nail to keep Rowan and the baby with her rather than give them up into the care system. That she'd become a mother as she reached for the baby and took her from Rowan's arms. As she looked for Rowan in a crowd of misguided fools intent only on death. As she had fought Lisette to keep her away from the children she wanted to poison.

'I can't do it alone,' she said.

They had lost each other for fifteen years and found each other against the odds. If he took her in his arms now, he'd never be able to let her go again. They could do it together. Make the impossible happen for themselves. Create a family.

'You won't be alone, Mia. You'll have the kids.' Instinctively, she turned her head to check again on Rowan and the baby. Her eyes, when they met his again, brimmed with unshed tears and unspoken words. That she loved him. That she knew he loved her too. That this was goodbye.

'I'm going to have a huge hit with this TV series,' she said. 'I'm going to win awards.'

'I never doubted it.'

'I was naive when I got here. I thought if I exposed Enoch it would be enough to stop him hurting other people.' Her face set. 'I should have tried harder. Not everyone here was evil. Most of them were lost souls.'

'That's not on you, Mia.' He sensed the presence of a police officer. They wanted Mia in the helicopter and off the island. 'I have to go,' she said. The back of her hand grazed his. Skin against skin. *I like them. Your patches are cool. Lend me some.* A thirteen-year-old boy comforting a twelve-year-old girl.

He watched as she walked away. Raising a hand across the distance to Rowan, who waved back. Rowan would be fine, he thought. Better than fine. The kid was resilient and brave. He had someone to love and nothing and no one was getting in the way of that. He deserved more than a life in the cold and a mother who didn't love and protect him. He deserved Mia.

'Mickey!' Her voice cut through the noise of the rotors and the wind and the police officer who stood by him, already asking him questions. And her face was the face of a girl who'd been his friend when he'd had nothing and no one. 'Don't disappear on me again, okay?' she called, walking backwards. 'Don't make me come find you, fool.' She blew him a kiss and turned back around. There was still time to go with her. Time to step into a different future. But he stayed where he was.

They debriefed North on the island and only then did they allow him to leave. It took another four hours to make it back into the centre of London. He got the news on the way.

North looked around the hotel suite. It had to be a mistake. Crossed wires. A miscommunication. Through the open doors from the living room into the bedroom, he could see Granny Po on the oversized bed, curled over on herself like an autumn leaf, her crumpled eyes squeezed tightly shut. She crooned in a low moan as if an animal in distress.

Was she in pain? Was she dying? But not even the prospect of death would reduce Granny Po to this. No, the old lady had a will of iron and he didn't think she could die. He thought she planned on living forever. Fate wouldn't have a say in it. Only one thing would reduce her to this. He felt a nerve twitch in his jaw.

North walked over to Fang's bedroom and pushed open the

door. The room was empty. Not a sparkle. Darkness settled over him.

'What happened?' he said, walking back through to the living room.

Seated at the dining table, Plug's face was bloodied and bruised. He was holding his ribs – five of which North knew to be broken. His eyes downcast, Plug seemed unable or unwilling to speak.

North crossed over to his friend. 'Where is she, Plug?'

Hone's voice came from the doorway into the corridor. 'She's gone, North. That's all you need to know. The Americans have her. And good riddance.'

Good riddance to Fang?

At the words, North moved so fast that Hone didn't see him coming. Taking hold of the other man's lapels, he dragged him into the living room, and slammed the door out into the corridor. He shoved his boss up against the wall, his bunched fists pushing into Hone's chest. The anger that had been building in him ever since he got the news reaching a head.

'What do you know, Hone?' He sensed Granny Po sit up in the bed, heard rather than saw her slide down from it. He'd witnessed nearly three hundred people kill themselves. He'd fought with a psychopath in a blazing building, and a violent gangster in the sea. And before any of that, he'd woken up in a bed filled with snakes and narrowly escaped falling to his death when his climbing rope was cut. The fact he'd had to jump from Tower Bridge to escape a lunatic seemed a lifetime ago. As he'd come up in the lift to the penthouse suite, he'd admitted to himself he could do with a break. Once he had checked back in with Linklater and told her he wasn't dead yet, maybe he'd go somewhere sunny. No islands though. If only this story about Fang wasn't true. If only she was safe with Granny Po and Plug in the hotel room – waiting to make some smart alec comment about how dumb he was.

Hone struggled to break free of North's hold, but North had the advantage of youth. He held the one-eyed man's hand down by his side so he couldn't reach into the pocket of his riding coat. North sensed no surprise. Violence was the reaction the one-eyed man had been ready for. He took a breath. It wouldn't do to lose his temper. 'Tell me all of it, or you're a dead man.' His voice was cold and eerily calm, but he didn't let go. The message was clear. He'd kill Hone if that's what it took. Fangfang was family. Hone, it turned out, was no one.

There was a pause as his boss appeared to weigh up the statement. He let out a sigh of impatience, before wrestling himself free of North's grip.

North let it happen. But he didn't step back.

'All right,' Hone said. 'But you aren't going to want to hear it. Fang gave us up. Gave you up.'

Incomprehension filled North's head. 'Not possible.' Instinctively, he moved away from Hone, who took the opportunity to head for a silver tray with a coffee pot and cups laid out on the low table in front of the brocade couch.

'They had her.' Hone allowed himself to sound exhausted. He sat down and poured himself a coffee. Was it hot? Was it cold? Who cared. Behind him, North noticed the London sky was a brooding charcoal grey. 'She'd fished in one of their ponds and they came for her. They made her an offer and she felt she had no choice. Sometimes the way this goes is if the Yanks make enough of a fuss, we give up the hacker, maybe in a week, maybe in a month. We close our eyes as they ship them out the country and throw them in a dark hole for however long they want. It happens more often than you think. This time though, because of who she was, who she worked for, they took a different approach and they were damn clever about it.'

The heat in him died away. Instead, an arctic cold anger flared and crackled somewhere inside North. Through past experience, he knew this was when he was at his most dangerous. Not when he was raging, but when he felt himself to be fully in control of his smallest emotion. When he felt his blood settle and fill with chips of ice. He watched the one-eyed man's face. 'She's not a freelance hacker nobody. She's Fangfang Yu and she's your agent.'

'Is she though?' The one-eyed man gave every impression of not caring. 'You're my agent. She's tech support at best. Child labour at worst.'

'She's under your protection,' North reminded him.

Hone made an almost imperceptible expression of disgust. 'You have to earn my protection.'

North had always known they were disposable to him, but he'd counted on the fact they were useful, had counted on the fact Fang was still a child and Hone would always do what he could for her. 'You're telling me you'd have let them take one of your agents.'

'I'm telling you we are none of us playing in a sandpit. She messed up big time and she knew the consequences. If she hadn't given us up, if she'd told me when they first came for her, I could have done something. But she cooperated. She betrayed her country. They knew what they were doing. It means she has no friends left to fight her corner. No one goes in to bat for a traitor. She belongs to them now.'

North considered shooting the one-eyed man dead where he sat. To see if it did at least ease the pain Hone was inflicting just a little. But he felt Granny Po's hand on his, and he glanced down at the wrinkled face. She was right next to him and she was shaking her head. There was something in her eyes. A warning. A plea. Hone didn't notice.

'Tell me where she is.'

'Why? There's nothing you can do for her. Do you understand? She panicked and she gave us up. Ironically, it's not all bad. Do you know why we got such speedy cooperation from the Yanks when we called them for backup for you on Murdo? Do you think that was down to the "special relationship"? Sure, the Americans helped us out. But thanks to Fang, they got a head start on shutting down their operation on Murdo. Everything Benson was doing there was sanctioned by the US Space Command and the Pentagon. They cleared every last paperclip from the place. If we'd been able to prove what was going on there, that would have given us leverage. The kind of leverage that could have forced the Americans into abandoning their programme. As it is – who knows what they'll do with their weapons. They could find another test site tomorrow. They could be slinging those laser weapons into space as we speak.

'If she ever comes home, and I mean – ever – we'll throw her in prison darker and deeper than the one the Yanks would have used. I'll make sure of it, and I'll turn the key in the lock myself. She's beyond the pale now, North. A traitor. Fangfang Yu is lost to all of us. That's the business we're in.' He stood up from the sofa with a heavy sigh. 'Let her go. I have.' And with that, he walked slowly towards the door. At the threshold, he turned back.

'Consider your next move carefully, North. You're still here and useful to me. You still have a job with me if you want it.'

He closed the door behind himself with a soft click that sounded far too loud in the sudden silence of the room.

North sat on the brocade couch with his head in his hands.

The fact Hone had helped Mia get on the island seemed irrelevant now. Who cared why? Hone's machinations were beyond him. He could sense Granny Po attempt and fail to drag something out from under the hotel bed. He went over to help. It was a massive box, tied with a huge red ribbon and wrapped in red shiny paper. He lifted it up onto the mattress – it was heavy. A gift tag hung off it. 'In case bridges remain a thing!' He'd never seen Fang's handwriting before. He'd presumed she could write. But she was permanently attached to a keyboard.

Granny Po delved in her knitting bag. There was a rustle of paper as she pushed something under his nose. It was an envelope, and his name was on it. He closed his eyes. He ripped it open. Hotel headed paper, which she must have taken from some leatherette binder left in the room. He closed his eyes. Did he want to read what she'd written? About what she'd done? If he never read it, none of it had happened and he could pretend everything was still good.

Hey Moron-person,

I figure you know it all by now (makes a change!). I have a cushy gig waiting and it'll be an adventure. I'm all right with how it's gone down. Ask Plug to take in Granny Po for me. She'll die of boredom if she goes back to Mum and I can't bring her with me. Whatever bollox Hone tells you, I didn't do anything that would be a problem for you. But I'm sorry about the drama – it's not how I wanted it.

Remember, bozo, violence can't always be your go-to. Use your head.

*(And don't go turning the world upside down looking for
me. It's all good, moron-person. Happy Christmas!)*

Ciao, Fang.

Plug loved Granny Po almost as much as she loved him. The
two of them would be fine. He hoped Plug's wife would be
cool with taking on a granny. Perhaps she'd just be so relieved
Plug was still alive, she'd be okay with it.

He stood up, crossed over to the bed, knelt down and tore off
the Christmas paper. As frightened as she must have been, with
typical generosity she had thought of a goodbye gift for him.
Had written him a letter rather than just disappearing. And
she was 'sorry about the drama'. The drama of being ensnared
by the Americans, abandoned by Hone, and shipped off to a
strange country to be exploited and blackmailed. The kid was
fifteen. She was a genius, but more than that. Fangfang Yu was
special. Brave and compassionate and loyal. And he had no
intention of letting her rot wherever it was the Americans had
her. One way or another, he was bringing Fang home.

Granny Po moved close to him. She knew what he was
thinking, and he knew what she wanted, and it wasn't to go
live with Plug above an undertaker's. He held her apart from
him, her arms like chicken bones. 'You can't come with me,
Granny Po.'

She let out a mew of protest. Deprived of Fang, she'd got
older, he thought. The wrinkles and furrows carved deeper in
the round face, the eyes glassy with unshed tears. 'Stay with
Plug. He needs you and I got this – trust me.'

His friend's gravelly tortured voice sounded as if it was being
hauled out of the depths of hell. Plug was blaming himself for
not fighting harder. For not getting the better of six CIA agents,

all of them armed. 'It's true Fang said the CIA were on to her. That's why she had to come to mine. But as a rule, that one-eyed scumbag is full of shit. We all know that. You aren't doing this on your own. Let's go get our girl, North,' Plug said.

'You have a family to think about, Plug.'

Plug snorted. 'Fang's my family. Granny Po's my family. You're my family.'

'What about the fact you're hurt, Granny Po is older than Time. And I have a bullet in my head,' North said. 'We make quite the team. Plus, without Fang, we have no brains trust any more.'

'You have me.' North hadn't heard Bald Paulie come in. He was smiling affably, his face shiny with good will, an enormous pull-along suitcase next to him.

'Damn right,' Plug said. 'Let's bring Fang home.'

North didn't reply. Because he didn't have to. He knew this much. In her letter, Fangfang Yu had said she was sorry for the drama. But Fang never took a step without knowing it was taking her where she wanted to go. And she didn't have it in her to commit treason. She'd rather die where she stood than betray a friend or a cause or her country. Which meant that whatever she had done or signed up to do had been done for a reason. And her survival – her freedom – wasn't it.

62

London

Eugenie walked her cocker spaniel every evening in St James's Park. It served as her own exercise as well as his, she thought. It was her thirty minutes of respite. She refused all calls, even those from the PM – especially those from the PM – and instead used the time for thinking. She often thought that it was the most valuable time of the day. No knocks on the door requesting her immediate attention at meetings, no briefing papers, no media firestorms. She grunted in ill temper, then, when she noticed the one-eyed man's presence on a park bench just down from the pavilion and overlooking the lake.

She didn't bother disguising her mood as she sat down along from him. It was as well she had worn her mac.

'You should have set up a meeting,' she said, taking out her e-cigarette and switching it on. Popcorn-scented. It tasted filthy but the smell of it reminded her of the half-term fair she used to attend with her sister.

Hone didn't respond. She had the urge to be rude to him. Prove who was the boss, but something about the man unsettled her so she contented herself with scowling at an old woman throwing bread into the water for the pelicans. 'Something of a shitshow on Murdo Island,' she said. 'I told you we should have sent in Special Forces. It was too much for one man.'

'I remember,' Hone said.

She reminded herself that she'd dealt with men like Hone all her life. They didn't intimidate her. Not even this one.

'The Americans aren't happy. They've pulled out of their lease for the station. They say they won't be needing the site again. That there's too much attention on the island.'

Hone reached into the pocket of the long riding coat and for a split second she thought he was reaching for a gun. Instead, he pulled out a packet of cigarettes. He didn't offer her one, which she told herself she was glad about because she would have put away the vape and taken it. *Never take sweeties from strange men*, she heard in her mother's voice. Never get in their car. Never believe a word they say. Not even as a child had she ever come close. She was altogether more canny than her sister ever was. She tossed the vape into the bin close by the bench and heard it knock against the side.

'It's a strategically advantageous spot ·when it comes to keeping an eye on the Russians,' Hone said. His voice considered but neutral. 'I'm told our own defence people can't wait to be in there again.'

Eugenie thought about the leverage that piece of rock miles away gave her. How grateful the chiefs of staff would be, but she was too astute to allow her face to show anything approximating satisfaction. 'Something of a silver lining,' she said.

Did she feel bad about any of it? Violent delights have violent ends, she reminded herself. Those who had joined the cult were fools and she had no time for fools. But she did allow herself to feel some sympathy for the grieving families they'd left behind. Families whose tearful interviews were all over the national media right now. And the government had already announced an inquiry into the whys and wherefores. Not that it would ascribe blame where it mattered.

'But we can at least take comfort in the fact the Narrow Yett is no longer a threat to national security,' Hone said. 'If indeed it ever was.'

Eugenie went very still.

'And Enoch is dead. His proverbial head in the proverbial box. That, at least, must make you happy.'

No one watching the one-eyed man in the riding coat and the woman with the spaniel sitting on the park bench would have thought they were doing anything other than passing the time of day, Eugenie thought.

'I have things to do.' She made to stand up, but he gestured at her to sit back down. She gazed at him. Desperate, suddenly, not to be anywhere near Hone. She sat back down and faced the lake, where the waterbirds swam this way and that.

'At first, I thought that's all this was. Some long game you were playing with the Americans. I asked myself if you were trying to keep them happy by clearing the Narrow Yett off the island and giving them the run of it. Bearing in mind what we know about what they were doing there, I thought you might have secured some quid pro quo on the satellite weapons. That British satellite interests up there would be unharmed. That we could call on them as well. That kind of thing.

'Then I thought maybe you wanted them to believe that. And that what you really wanted wasn't just the Narrow Yett gone, but the Americans with them. That it was a house of cards and you thought if you shone enough light on the island, the Americans would get "frit" and leave. Still, that would only ever be a long shot. You weren't to know what Enoch would do. How many deaths there would be.

'Eventually, what I realized was that you never really cared about Murdo's strategic significance at all, did you, Eugenie?'

She noted his use of her first name. One he had never used

before, she realized. The usage making them seem closer than they were, that their relationship was more intimate than it truly was.

'We'd suffered the Americans on that island for decades,' Hone said. 'At first, we even wanted them there. But even when we didn't, we let it go. All of a sudden, though, I'm hearing the Americans are an inconvenience. They're making the island a target with whatever it is they're doing. They're up to no good. All of which was true. And yet not the whole truth…'

Eugenie let out a sigh that felt as if she had been holding it inside of her for years. Perhaps because she *had* been holding it inside of her for years. 'Cut to the chase, Hone. Enoch Fraser was an evil predatory monster. He was a psychopath and his actions on that island prove it.'

Hone let go a plume of cigarette smoke – she hadn't seen him light his cigarette, she realized. How did he do that? But that was spies for you. Sneaky bastards. 'I don't take issue with that,' he said.

'My sister Maria was brilliant.' Eugenie made eye contact with Hone for the first time in the conversation. 'My father knew the kind of man he was the first moment he met him. He warned her. I warned her. But she ignored us. She was in thrall to him.

'You saw what he did to those people on that island. What he convinced them to do. I never underestimated him. He was nothing to look at but when he turned the full glare of his attention on you, it was as if you and he were the only people to exist in the world.' She hesitated. 'She was so happy. Her business was booming. Enoch appeared to be both brilliant and devoted. I was beginning to wonder if we'd misjudged him. And then we got the call. She was dead. They'd gone skiing – off-piste in the Italian Alps, and there'd been a terrible accident.

But we knew it wasn't an accident. She left him everything. We put our own private investigators on it but they could prove nothing. We fought him tooth and nail in the boardroom and eventually got him out of my sister's company, but it cost us every penny we had. It broke my father's health. My mother died a year after Maria. Enoch forgot her and he forgot us. He had her money and that was all that mattered for a while. Except he needed more than money. He needed control. I kept track of him. That ridiculous book he wrote. Kept track of his cult. You're right. He got invited to a Downing Street reception. The PM's wife, you know, loved him...' Eugenie rolled her eyes. 'But I was the one who wanted him there. I wanted to look into his eyes and see who he was. I saw the evil. I thought he might recognize me, but he didn't. He was focused on the PM, on the PM's wife, intoxicated by his own importance. I was irrelevant. The next day, I put the Inland Revenue on to him and they might have pinned him down. I figured he would never play with a straight bat when it came to money. But then Gaffney came along and Hoorah! Finally, my chance to shut him down. To destroy him like he destroyed my sister.'

She stopped talking. Her face was flushed.

'I don't care if the story comes out. He's dead and gone, that blaggard Drummond with him, which is all that matters.'

'It doesn't have to come out. I don't see why it should. After all, if it did, questions would be asked whether the authorities were heavy handed for a reason. I can't see that being a comfortable experience for you.'

She gave him a long, even look. That's how these things worked. A favour here. The right word there. She would owe this creature of the shadows and when he wanted something – and he would want something – she would have little choice in the matter.

Eugenie stood, her little dog dancing at her feet. Sunlight glinted off the water. The pelicans preened themselves. A stray fact popped into her head – that the pelicans here were originally a gift from the Russian ambassador to Charles II. Did Hone know that? He knew most everything else, she thought.

But by the time she turned to ask him, he had already gone.

63

An undisclosed location

Outside, the air was baking hot. Inside, the air con was turned too low but Fang figured they didn't want the machines playing up. Shivering, she made a mental note to wear more layers tomorrow. Or find a blanket and wrap herself in it. Roxie pointed with her tiny hand to a space along the row. Every head in the place was bent over their monitors and nobody looked up. The only sound in the place, frantic typing. Fang yawned. The jet lag was killing. Roxie had dragged her bodily from bed this morning before throwing her into a cold shower. Plus, she was starving, but she refused to give Roxie the satisfaction of saying there was no time for breakfast.

She took her place. And allowed her eyes to slide side to side as she took in her work companions. She opened her mouth to ask about food as much as anything else, but a voice hissed at her. 'Say nothing. They listen.'

'Welcome to the dark side, Miho.' The words appeared and then disappeared on her screen before she even had time to blink. They knew her hacker name. Had the people here helped devise the trap that had closed around her and swallowed her up? Or were they victims as much as she was? Could she trust them? She didn't think so. She didn't think she could trust anyone but herself.

She missed Granny Po. Missed North. Missed Plug. She even missed her mother. And she wondered what they were all doing. Was Plug looking after Granny Po? Had North liked her going-away gift? She wondered how long he would last without her telling him what to do.

Behind her, she heard the squeak of a shoe as a thin man she hadn't noticed before headed towards Roxie. Fang glanced down – brown leather brogues – then up. Mouse. 'Welcome to America, kid,' he said, and kept walking. Benson Corp was part of this. Of course it was. Benson and the US defence establishment were tight. It made sense they would have cyber interests. She knew Roxie and she knew Mouse and she knew both of them were killers. Roxie, because her instinct told her so. Mouse, because she had caught him in the act. Who were the rest of these people? *And what was this place?* And for the first time in her life, Fang wondered if she had made a mistake.

64

London

At weekends, Hone lived in a Georgian stone-built house in a picturesque Derbyshire village where his neighbours believed he was 'something in the City'.

During the week, he lived in an apartment overlooking the Thames where his neighbours never saw him. Ironically, it wasn't far from Tower Bridge. This particular night, he had switched on the lights and gone straight to the bottles of Macallan whisky. His hand hovered over the eighteen-year-old and the Engima, before settling on the Enigma. It had been a long, exhausting day – he deserved some joy out of it. News outlets had gone crazy over the leak of an interview with an American bioscientist called Dr Libby Thornton, apparently admitting an aerosol neurotoxic bioweapon experiment on Murdo. Benson's lawyers had issued a full and categoric denial of the bioweapon claims and would only admit to a minor leak of rocket fuel on the island. Dr Thornton, they said, was a victim of fake news. She was, as yet, unavailable for comment. The voice of the interviewer had been that of a young American man who had claimed to be a senatorial aide. The aide denied all knowledge of contact with Dr Thornton and claimed himself to be mystified as to the real identity of the interviewer. The latest FT.com headline Hone had seen read: 'Hoaxer exposes Benson's bioweapons work. Company market value plummets 25%.' While the *Daily Mail* asked: 'Were Doomsday Loonies Poisoned by Yanks?'

Hone knew exactly who the hoaxer was. And was surprised only that Fang had managed to expose the bioweapons before they shipped her out of the country. They might suspect she was responsible, but they'd struggle to prove it.

He calculated that he had four hours' sleep before the car came for him at five the next morning, but it would be enough, and he would feel neither better nor worse than he always felt.

The whisky was in his mouth and the Glencairn glass still at his lips when he realized the French windows were open, and worse than that, there was a figure out on the balcony. He turned around slowly, and as he did, he considered the lead pipe he had secreted in the lining of the riding coat he still wore, the gun holstered on his hip, the glass in his hand and the knife within reach if he could only put down the glass. He had options. None of them great. But it could be worse, because at least he was still alive.

The window creaked open and a voice came out of the darkness. 'Do you know how many men I've killed, Hone?'

Hone had a rough idea. More than he had, which was saying something.

'One more won't bother me.'

'Doesn't that depend on who it is?' Hone gave up on the idea of throwing the glass and reaching for the gun. Or even trying for the door. North was too fast for any of it.

'Tell me where Fang is and what you did.'

'You read the letter...'

From the balcony, North fired a bullet and it shattered the glass in Hone's hand. Shards flew everywhere, chips pitting and stinging skin already damp with the whisky – the smell of it coming up from his coat. The gun had a silencer attached but even so, it was noisy.

'That was risky. My neighbours will wonder what's happening.'

As he said it, behind North a firework exploded in the night sky, a chrysanthemum of blood red light and then another, this one in gold. And another – this time in blue.

'Is that right.'

Hone shook his head in frustration. He'd been so close to pulling it off without North being any the wiser. In the hotel, he'd thought it had worked.

'You're making a mist...' From the half-lit darkness, North fired another bullet and this one grazed the tip of Hone's ear. He let out a yelp and some Anglo-Saxon. His fingers touched his ear and they came away bloody. North wasn't kidding. But then he should have expected as much.

'It's exactly as I said. The Americans came for her. They'd caught her nose deep in D'Urberville and once they tracked it back they had her fingerprints all over various of their defence sites. They owned her. They told her she was on a plane and in lock-up for the next however many decades. But they made her an offer – if she gave us up on D'Urberville and the Narrow Yett thing, they would bring her across and set her up over there. They wanted her as a double agent, temporarily at least. Then working for them full time.'

'It's eminently believable.'

'It's what happened.' Hone allowed himself to feel a moment of sweet relief.

'But is it everything that happened?'

'I don't know what you mean, North.'

'She'd never do it. Never in a million years. That's how I know you're lying. That and the fact you couldn't walk a straight line, Hone. You're the definition of a crooked man in a crooked house. And I should have known better than to ever trust you.'

Hone let out what could have passed for a laugh. 'Have you ever made life easy for yourself, North?' He shrugged. 'Okay, you're right – she played them. She gave them a memory key with information about the operation on it. Information she figured they already had. It gave her a backdoor into their systems and she told me what was going on. We decided to turn it round on them. I made a show of firing her. Of pushing her out into the cold. She's over there, but she's still working for us.'

'So you and she "decided to turn it round on them"? You, the slipperiest bastard I have ever met, and your teenage agent "decided" to put on a show.' North moved to the threshold but didn't cross it. 'Now tell me the truth, the whole truth, or I'm going to put the next bullet in the place where your heart should be. Don't think for a moment I won't do it. When it comes right down to it, it turns out that you're nothing to me.'

There was a moment's pause that might have been surprise. Hone had several fallbacks in the event of something like this happening, but North was closer to the truth, faster than he had expected. He sighed.

'All right. Since 2014 and 2015 when the Russians hacked Ukraine's power grid, the Americans have been going gangsta on cyber warfare. They have three Cyber Command factories set up across the US stuffed full of their best hackers and tech brains. Officially, they're identifying and attempting to mitigate threats. Unofficially, it's an offensive game for them as much as defensive. We've been looking for our way in ever since we figured out what they were up to, and Fang is it. She's brilliant and she's vulnerable because of her age. We figured if we dangled her out there, they'd bite, but we couldn't afford to make it too obvious.'

'They're our allies.'

'Sure. And information will help keep them that way. Fang is going to have unprecedented access to their cyber warfare strategy. We're going to know every move they make, every program they write, every target in every country. We're going to know who she's working alongside and who's in charge of her. She's going to be in a position to recruit some of them. This is a long game. With Fang in place, we can shape what they do and how they do it. They'll have no cyber secrets left to them. To state the obvious, we need it for our own defences. If we aren't proactive on this, the Americans can hold us to ransom as much as China or Russia or Iran. Loyalties change. They can shut down our hospitals, our banking system, the police, government itself. You have to see the sense in that.'

Sirens. The neighbours had indeed called the police. Perhaps because of the firework display, which would mean the police spent time on the ground looking for the miscreants rather than in the apartment block, breaking down Hone's door before he was murdered where he stood.

Hone was silent, watching as North did the maths.

'And you pulled this together on the fly? I don't believe you. You never cared about the Narrow Yett, did you, Hone?'

Hone wanted a cigarette, badly. But if he went for one, he risked North shooting him. And he wouldn't ask North for his permission. North was his subordinate, after all. Or at least he used to be. What he was to him now, he didn't want to think about.

'We didn't know what the Americans had done on the island. We weren't to know they'd used the cult as guinea pigs.

We couldn't anticipate the extent of Gaffney's paranoia and what he was capable of. Local police let us know that Gaffney was off the island. Being their finance director, he was of some interest. I thought we'd be able to persuade him to cooperate. I'd have brought you in and sent you over at that point. But I underestimated how troubled he was, how messed up by Benson's neurotoxic aerosol.'

'And Mia?'

'We knew Mia had her own arguments with Enoch and the Narrow Yett and that justice hadn't served her well. We arranged a meet and had a conversation. We advised on her approach and asked nothing in return. It was up to her to persuade them to give her that access, though, and she did a good job.'

'And you did that not out of the goodness of your heart, but because you knew with Mia on the island I would have to go.'

'It was always the plan to get you on that island. Gaffney speeded things up is all.'

The sirens were closer still. Outside the apartment building now, he thought. Were they close enough to hear him if he shouted for help? Possibly. But he'd have hit the floor and breathed his last before help came.

'None of this was really about the cult, was it.'

'We had no idea how bad things there were, so no. It wasn't about the cult.'

'You got Mia onto the island in order to make sure I went up there. Why?'

'The Americans had leased the rocket checking site since after the Second World War. And they are, of course, entirely right to exercise a degree of paranoia about Russia. Plus, they like their faraway places with plenty of deniability. If you were out on the island, we figured you'd certainly come across them at some point and do due diligence. You'd ask Fang to check them out. She'd start digging into D'Urberville and Benson, which would take her further into the American defence sector. That's what the end game was.'

'This whole thing has been about Fang?'

Hone nodded.

'You've been planning to use Fang to infiltrate the Americans' own security system for some time then?' North said.

Hone looked thoughtful.

'As soon as I met Fang, I started thinking about how we could use her best. She's brilliant, but I needed to know she was steady under pressure. That she could cope on her own out there. I realize she's useful working with you, but frankly she's wasted doing what she does for you, when she's capable of greatness.'

'I don't understand. Last year, when Lilith took her hostage, you'd have let her die.'

'I told you then I wouldn't have let that happen and I meant it. Right this second, Fangfang Yu is the most valuable asset the British government has in play anywhere in the world. I'm proud of her.' Hone's wrist gave an almost imperceptible flick and the lead pipe dropped into the cup of his hand. He was too far away for it to reach North if he swung it, but he considered throwing it with as much force as he could muster. It could act as a distraction at least. He could try to take cover and reach for his gun.

'Don't!' North warned, and Hone let go of the bar. It chimed as it hit the floor. North was good. As good as anyone he had ever come across. And a moment of regret passed through Hone. Brick by careful brick, he'd built up North's trust and that was a dangerous thing to do. When you betrayed North, punishment was swift and brutal, Hone had seen the bodies to prove it. He wondered if his own time was up. There'd been difficult moments before. Moments like these, face to face with dangerous men. But he wasn't fooling himself. This was closer to the edge than he had ever come.

He had to proffer the truth. Some of it, at least. 'We never expected it to unwind the way it did on Murdo. We certainly had no idea about the bioweapons. It's messy but it's still a win. Downing Street is happy, the cult is gone, your friend Mia gets a hit TV show and a ready-made family, we acquired valuable information on the Americans' covert military operations in space. They're embarrassed about their bioweapon programme. But most important of all, we get our asset into the heart of the US cyber warfare system and Fang gets to be a hero. Don't think she doesn't want that. Everyone's happy and that includes Fangfang Yu.'

'Happiness is overrated.'

Hone couldn't argue with that.

'She was always in play and you weren't straight with her. She never knew that. She still doesn't. Her decision to cooperate with you wasn't an informed one. You don't really know what she wants and you care even less. Men and women died on Murdo. Innocent bystanders died on Tower Bridge. Fang might die if they realize what she's up to. All because of this operation – because of you.'

'Do you know how many UK citizens will die if hostile agents bring down a plane over a city? If they crash it into a nuclear power plant?'

'She's a kid.'

'She's an agent before she's a kid.'

'Now she's an agent?'

'She was always an agent.'

'Where is she?'

'I can't tell you.'

Outside the window, the firework display was coming to its climax. Bangs and shrieks and spiralling lights of red and blue and white and yellow gold.

'Then this conversation is over.' North lifted his hand a fraction and the barrel of the gun lifted too. Hone wasn't frightened because that wasn't how he was built. He was a rational man. Some information was worth dying for. But not this, because North would find out another way. This was sliding out of his control. He held up a hand.

'All right, New Mexico. Not far out of Albuquerque. In what used to be a meat-processing factory. They house their cyber warriors – and yes, that is what they call them – in a specially built motel complex not far away. But you can't get her out of there. It's too late. If the Americans realize this was a sting, it's over for her. If they even get a whiff of it, it's over for her. One way or another. Do you understand what I'm saying, North? They're never letting her go.'

'Is that right?' North rose, and the handsome, thuggish face was cold. 'We'll see about that. By the way, to state the bleeding obvious, Hone, you and I are done.'

To his surprise, Hone felt regret. North was young but gifted. Hone had given him a purpose and he had exploited him every second of every day. But there were always difficult choices. He'd had to play the Fang card when the time came. It was his duty and he was doing it. If he had to pay the price for that, so be it. They all paid a price one way or another.

North gestured with his gun for Hone to turn around, and Hone braced for a bullet in the head. For shattering pain and then nothing. He wouldn't get to be as lucky as North, he thought.

Instead, he heard the sound of engines starting up. A smell

like kerosene. An intense wall of heat. On the railings of his balcony, North stood, fire flaming from his outstretched arms.

'If I ever see you again, Hone. If you ever come near me or any of mine...' North's voice was loud to be heard over the roar of the engines. '... you're a dead man.'

With a jet-engine roar, he took off from the balcony, lifting up near vertically before swooping and swerving through the explosions of blossoming colour and light up into the starry London sky, and across the rippling Thames. Like an avenging angel, Hone thought. Despite himself, he shivered. And as he did, his one-eyed gaze fell on the mirror by the window. *See You in Hell,* scrawled over the reflected fireworks.

Acknowledgements

Thanks to editor extraordinaire, Laura Palmer, who not only makes my books infinitely better but who is a pleasure to work alongside. Also the brilliant Head of Zeus team including Peyton Stableford (who is across everything!); Anna Nightingale; Jenni Davis, and the proofreader; Dan Groenewald and Lottie Chase in sales; Andrew Knowles and Paige Harris in marketing/social media; as well as PR supremo Sophie Ransom. Also Jessie Price and Matt Bray in the art team and @Kidethic for the re-brand. Great job, guys.

My spectacular agent, Jon Wood, who inspires and encourages at all points. As well as Safae El-Ouahabi at RCW.

As ever, I have been kept right by a small battalion of experts/ pals. (Any mistakes in translation are, of course, mine.) My experts include: Derick Hatton and Andy Tennant who advised on my fight scenes; John Tickner (helicopters); Helen Le Fevre (medical); Fred Cook (who devised an entire crossword for this book and who is patently a genius); Eric Pinkerton and Helen Chappell (IT); John Woodman (finance); Joanne Robertson and Brian Reilly (Scotland); and Ian Sedgwick (electrics). Also Dr Bleddyn Bowen of the University of Leicester (space and satellites) and Dr Malcolm Bates (legal). North's original neuroscience courtesy of Eamon McCrory and Philip McGill.

Tim Pedley for his time and subbing expertise. Similarly, Angela Barnes, my first reader. Also Natalya Sharp and Ella Paul.

My research ranged far and wide, but I found particularly interesting a feature on www.medicalnewstoday.com ('Mass Hysteria: An epidemic of the mind?'). Other texts I consulted this time round include: *Encyclopedic Handbook of Cults in America* by J. Gordon Melton; *They Shall Take Up Serpents* by Weston La Barre; *Cults in Context* edited by Lorne L. Dawson; *The Branch Davidians of Waco* by Kenneth G. C. Newport. (Professor Douglas Davies of Durham University provided a reading list.) Also, *St Kilda: The Last and Outmost Isle* by Angela Gannon and George Geddes. Space/defence news came from all kinds of places. In other words, God bless the internet.

Lucy Sampson who kindly allowed me to borrow her name for one of my very favourite characters in the book. Lucy won a Northern Crime Syndicate contest to have a character named after her. *You waited long enough, Lucy. Hope it was worth it.* And while we are talking Northern crime writing, a hat tip to my talented mates in the NCS – Fiona Erskine, Trevor Wood, Adam Peacock, Chris McGeorge, Robert Scragg, Rob Parker, and Dan Stubbings, who are a constant source of friendship, laughter and support.

Fiona Sharp and the top team at Durham Waterstones for the outstanding job they do supporting local writers; my pals in the blogging community, in particular Anne Cater, Mary Picken and Gordon McGhie. Also Karen Robinson, Samantha Brownley, and, of course, our brilliant North East libraries. Karen Sullivan remains my go-to for all things books, and thanks to the phenomenal LJ Ross who generously gave me a blurb for *Curse the Day* (a blurb which now adorns the cover of *Sleep When You're Dead*). Jacky Collins and Frances Moon Walker – you are shiny stars.

And to those not yet mentioned but who keep my show on the road. As ever, Sophie Atkinson, Sue Brooks (my other crossword genius) and Andrew Macdonald, Tim Burt, David Paul and Clare Grant, Fiona Ward and Adam Lambert, Allison and Ian Joynson, Marie Sedgwick and Michael Reilly-Cooper.

And, of course, my husband of too many years to count, my children, my children's partners, and my mother.

About the Author

JUDE O'REILLY is the author of *Wife in the North* – a top-three *Sunday Times* bestseller and BBC Radio 4 Book of the Week – *A Year of Doing Good* and the Michael North action-adventure thrillers. Jude is a former senior journalist with *The Sunday Times* and a former political producer with BBC 2's *Newsnight* and ITN's *Channel 4 News*. Her Michael North series has been praised by bestselling thriller writers around the globe. She currently lives in Durham with her family and a very noisy poodle.